Glow Discharge Optical Emission Spectroscopy
A Practical Guide

RSC Analytical Spectroscopy Monographs

Series Editor: N. W. Barnett, *Deakin University, Australia*
Advisory Panel: F. Adams, *Universitaire Instelling Antwerp, Wirijk, Belgium*;
M.J. Adams, *RMIT University, Melbourne, Australia*; R.F. Browner, *Georgia
Institute of Technology, Atlanta, Georgia, USA*; J.M. Chalmers, *VS Consulting,
Stokesley, UK*; B. Chase, *DuPont Central Research, Wilmington, Delaware,
USA*; M.S. Cresser, *University of York, UK*; J. Monaghan, *University of
Edinburgh, UK*; A. Sanz Medel, *Universidad de Oviedo, Spain*; R.D. Snook,
UMIST, UK

The series aims to provide a tutorial approach to the use of spectrometric and
spectroscopic measurement techniques in analytical science, providing guidance
and advice to individuals on a day-to-day basis during the course of their work
with the emphasis on important practical aspects of the subject.

Recent titles:

Industrial Analysis with Vibrational Spectroscopy, by John M. Chalmers, *ICI
Research & Technology, Wilton, UK*; Geoffrey Dent, *Zeneca Specialities,
Blackley, UK*

Ionization Methods in Organic Mass Spectrometry, by Alison E. Ashcroft,
formerly Micromass UK Ltd, Altrincham, UK; now University of Leeds, UK

Quantitative Millimetre Wavelength Spectrometry, by John F. Alder, *UMIST,
Manchester, UK* and John G. Baker, *University of Manchester, UK*

How to obtain future titles on publication

A standing order plan is available for this series. A standing order will bring
delivery of each new volume immediately on publication. For further information,
please write to:

Sales and Customer Care, Royal Society of Chemistry, Thomas Graham House,
Science Park, Milton Road, Cambridge, CB4 0WF, UK

Telephone: +44(0) 1223 432360
E-mail: sales@rsc.org

RSC
ANALYTICAL
SPECTROSCOPY
MONOGRAPHS

Glow Discharge Optical Emission Spectroscopy

A Practical Guide

Thomas Nelis
Le Village, 26150 Ste Croix, France

Richard Payling
*Department of Physics, The University of Newcastle,
NSW 2308, Australia*

RS•C

advancing the chemical sciences

ISBN 0-85404-521-X

Published by The Royal Society of Chemistry,
Thomas Graham House, Science Park, Milton Road, Cambridge CB4 0WF, UK

Registered Charity Number 207890

For further information see our web site at www.rsc.org

Typeset in 10/12pt. Times by TechBooks, New Delhi, India
Printed by Athenaeum Press Ltd, Gateshead, Tyne & Wear, UK

Preface

Glow discharge optical emission spectroscopy (GDOES) is an essential technique for the direct analysis of bulk solids, elemental surface analysis and the depth profiling of thin films and industrial coatings. This book is designed for all those using or managing GDOES instruments and for those who would like to know more about the technique from a hands-on perspective. It will also aid those considering the purchase of a GDOES instrument, or those using GDOES results, to understand in detail how the technique works and what is involved in maintaining the instrument and achieving high-quality results.

The book is designed as if the reader has just sat down at their instrument. It begins with checking whether the instrument is on and working correctly, then deals with sample preparation, method creation and optimisation of the instrument and source settings, then calibration and drift correction if necessary, followed by analysis (bulk or depth profiling) and the interpretation and presentation of results. Then follow additional chapters on the theory of GDOES, line selection, and trouble shooting.

The book may be read from cover to cover or the reader may prefer to skip to individual chapters of immediate interest. The theory chapter, for example, is largely self-contained and presents an up-to-date account of the glow discharge plasma, sputtering and quantitative analysis. Several chapters deal with calibration, probably the most challenging area for GDOES operators, including all aspects of calibration: the number and selection of calibration samples, the choice and optimisation of the calibration function, validation, and the calculation of uncertainties. The descriptions are not aimed at particular instruments or manufacturers and are therefore designed for use by everyone working in the field.

Throughout the book, we have endeavoured to follow ISO and IUPAC guidelines on procedures and nomenclature. At times these recommendations are crucial to analysis, as is the distinction between precision and accuracy. At other times, they are in conflict with common usage in GDOES. Where there is no reasonable ambiguity, we have opted for common usage. Hence, we have used content or composition, in units of mass% or atomic%, throughout in place of the more familiar concentration, now reserved for mass per volume, g L^{-1}. But we have continued to use voltage instead of potential difference when speaking about glow discharge source operation.

We would like to thank the series editor, Neil Barnett, for his suggestion and invitation to write this book, and for his useful comments on the draft manuscript.

We are indebted to our GDOES colleagues, Max Aeberhard, Philippe Belenguer, Arne Bengtson, Olivier Bonnot, Patrick Chapon, Volker Hoffmann, Paul Kunigonis, Mike Winchester, and Zdenek Weiss, for their many helpful comments on the draft manuscript.

We are grateful to Max Aeberhard and Johann Michler, EMPA, Switzerland, for access to their GDOES data; Cyril Autier, Jobin-Yvon, France, Karl Crener and Alain Jardin, Certech, Belgium, Stephan Kuypers, Vito, Belgium, and Chris Xhoffer, OCAS, Belgium, for the kind use of their sputtering rate data; and Philippe Belenguer, Philippe Guillot, and Laurent Therese, CPAT-UPS, Toulouse, France, for the use of their RF data on plasma potentials and currents. We thank Patrick Chapon, Jobin-Yvon, France, for permission to show his small sample device; David Guisbert, Quality Associates Metallurgical Services, Niles, Michigan, USA, for permission to show his cast iron calibration; Volker Hoffmann, IFW Dresden, Germany, for permission to show his powder-mounting device; Ken Marcus, Clemson University, South Carolina, USA, for permission to show his sputter-cratered, artificial hip joint sample; Ken Shimizu, Keio University, Japan, for use of his sequential anodised aluminium sample; and Arne Bengtson, SIMR, Sweden, Annemie Bogaerts and Renaat Gijbels, University of Antwerp, Belgium, Mike Winchester, NIST, USA, Zdenek Weiss, Leco, Czech Republic, and many others for permission to use their published results.

Finally, RP acknowledges the expert assistance of his wife Barbara Hoving in setting up the presentation style of the manuscript, and for her abiding support.

To dear colleagues Max Aeberhard and Paul Kunigonis

Contents

CHAPTER 1

Preparation and Maintenance of Instrument

There are several different designs of GDOES instruments available on the market and they range in size from small bench-top units to large laboratory systems, depending mainly on the focal length of the spectrometer. The largest GDOES instrument ever built, one of the first but no longer made, had a focal length of 2.5 m. Today, spectrometers vary between 150 mm and 1 m in focal length. Despite these variations in size, they all have many features in common (shown in block form in Figure 1.1):

- glow discharge source
- optical spectrometer(s)
- vacuum pump(s)
- electronics
- computer and printer
- services: electricity, ultra-high purity argon

Some have additional services: water cooling, high purity nitrogen (for spectrometer), compressed air (for pneumatic valves) and a computer network/Internet connection.

Before beginning work on an instrument, the first thing to do is to check that each of these components is up and running. Walk around the instrument. Check that the area is tidy and the cables are connected. Check the services are working: electricity, gas(es) and cooling water. If necessary, check that the vacuum pump(s) is working. Check that the GD source is clean and ready and the computer is on. Run a common bulk sample for several minutes to 'condition' the source, *i.e.* warm and clean the source through normal operation. If necessary, check the alignment of the spectrometer(s). The instrument should now be ready.

Glow discharge spectrometers, like most analytical instruments, need regular maintenance and preparation for measurements. So, besides this initial inspection, consider regular maintenance checks. The purpose of maintenance is at least two-fold:

- to avoid serious trouble with the instrument and
- to avoid bad/wrong measurements

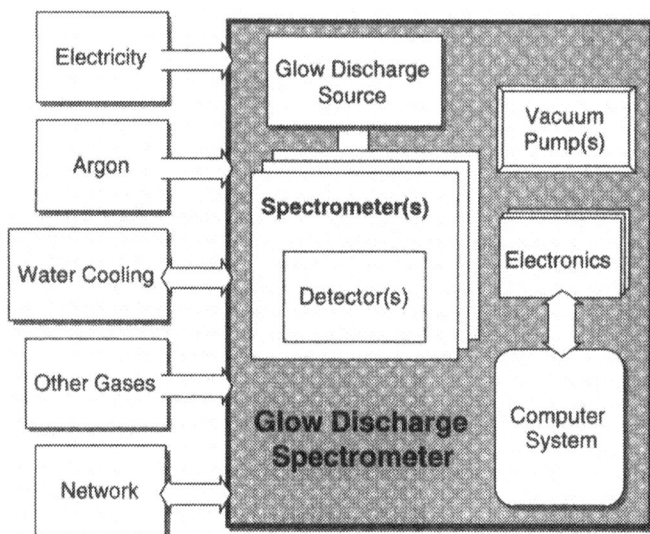

Figure 1.1 *Block diagram of GDOES instrument*

Checking the instrument before starting the analysis not only avoids trouble during the measurement, but also helps the operator to concentrate on the task of getting excellent results, as it liberates the mind from constant worries such as:

- Is my computer system stable?
- Is there enough argon?
- Do we have a leak in the source?

It is beyond the scope of this book to give detailed instructions for the maintenance of all instruments currently available on the market. For details on routine maintenance specific to your instrument, refer to the user's manual supplied with your instrument. In the following, we will provide general advice on maintenance procedures that will help save time in the long run and improve the overall quality of the analyses.

1 Before Starting Measurements

We will now consider preparation in more detail. Before starting to operate your instrument, make sure the computer and the software are in stable condition. This implies:

- if necessary, restart the system
- check the available space on the hard disk regularly
- archive the data you definitely do not want to lose

Checking the different supplies, such as gas, water and electricity, before starting analysis should become a habit. The argon supply pressure must be in the range specified by the instrument manufacturer. A different argon supply pressure may seriously affect the performance of the argon pressure regulation in the discharge chamber. When using gas bottles, be aware that the last 10% of the bottle may no longer meet the quality required. When replacing argon bottles, avoid getting air into the argon gas line.

Make sure the cooling system(s) is operating properly:

- check the water supply for open (non-return) systems or
- check the level of the cooling liquid for closed (recirculating) systems
- check for particles in the closed cooling liquid; if necessary replace the liquid

Make sure no power failure has occurred since you last operated the instrument, as this may seriously affect the condition of the instrument. For example, a long time without power may have affected the thermal regulation of the spectrometer, so the instrument may need time to stabilise. A common guide for larger, heated spectrometers is 1 hour stabilising time, up to a maximum of 8 hours, for each hour without electrical power.

When the instrument has not been used for some time, it is generally recommended to run a few dummy samples, before optimal stability can be achieved. This effect may be more or less important, depending on the design of the instrument.

2 Routine Checks

Perform regular routine analyses on samples of known chemical composition. If your instrument is used for depth profiling, these samples should include homogeneous reference materials as well as coated materials (see, *e.g.* Figure 1.2).

Figure 1.2 *Typical samples used for routine instrument checks: recal steel sample rich in C, S and P for checking window cleanliness (left), zinc phosphate-coated steel for O sensitivity (top) and galvanised steel for sputtering rate (bottom) (scale in cm)*

The purpose of these tests is to assess, verify and finally establish in written or recorded form the current performance of the instrument. They also help to detect possible malfunctions of the instrument, prior to analysis. Another result of routine checks on samples of known composition is the possibility of measuring the long-term precision and accuracy of the instrument for a given application, and the frequency required for drift corrections.

For bulk analysis, we suggest you analyse two different homogeneous samples, one a low-alloyed material and the other a high-alloyed material. The chosen materials should have a large number of specified elements. In particular, the elements contained in air, namely N, O, C and H, should be checked to make sure no air leak has occurred, as only a small amount of air in the discharge chamber can drastically change the nature of the glow discharge plasma, for all elements.

Analysing a coated sample allows you to monitor the sputtering rate and depth resolution. The sample used for routine checks should therefore be homogeneous in coating thickness and should have clearly defined interfaces between the different layers.

3   Monitoring Key Parameters Over Time

To monitor the performance of the instrument and to detect malfunctions before they become significant in the analytical results, the following key parameters should be monitored over time:

- intensities
- source impedance

An example is shown in Figure 1.3. These charts allow quick troubleshooting and can be used to plan preventative maintenance, thereby reducing unscheduled downtime.

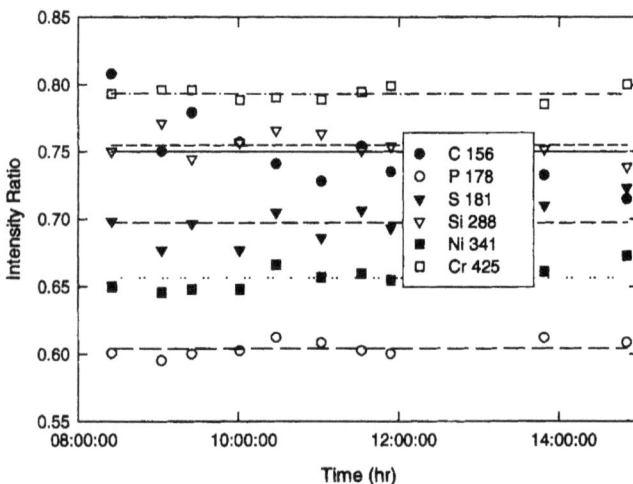

Figure 1.3   *Run chart for changes in intensity over time: here a cast iron sample is measured repeatedly over several hours, with intensities ratioed to iron. This type of measurement could be repeated over days, weeks or months, to see long-term trends*

Monitoring the intensities of the elements, N, O, C and H, over time, if they increase significantly or suddenly, can help detect air leaks in the argon supply system. All the wavelengths shown in this book are in nanometre (nm).

Monitoring the impedance of the GD source for a given sample can be used to record the reproducibility of the plasma, *i.e.* of the sputtering and excitation processes. In particular, it can be used to monitor the pressure regulation and any changes in the anode-to-sample gap over time, with the eventual need to replace the anode if the gap becomes too large. Changes in source impedance can be monitored, in RF operation, at fixed applied power and pressure by recording changes in the DC bias voltage or applied voltage over time. In DC operation, the voltage can be recorded at fixed current and pressure.

4 Anode-to-Sample Gap

The anode-to-sample gap is one parameter that has generally been neglected in most work on glow discharge spectroscopy. Though it is not so difficult to measure, no manufacturer currently provides a routine way of doing it. Despite this negligence, the anode-to-sample gap is a crucial parameter for reproducible GDOES data. Increasing the anode-to-sample gap increases the source impedance. This change in source impedance then affects the relationship between current and voltage in the plasma. The effect is shown in Figure 1.4 for a large change in anode gap from 0.12 mm to 0.30 mm. Such a change, if undetected, would have dramatic effects on the validity of calibration curves and the accuracy of analysis.

An anode-to-sample gap between 0.08 mm and 0.2 mm is generally recommended. Small gaps are usually beneficial for improving depth resolution; larger gaps allow running the discharge for longer times without creating a short circuit between the anode and the sample. Whatever gap is chosen by the manufacturer or the operator,

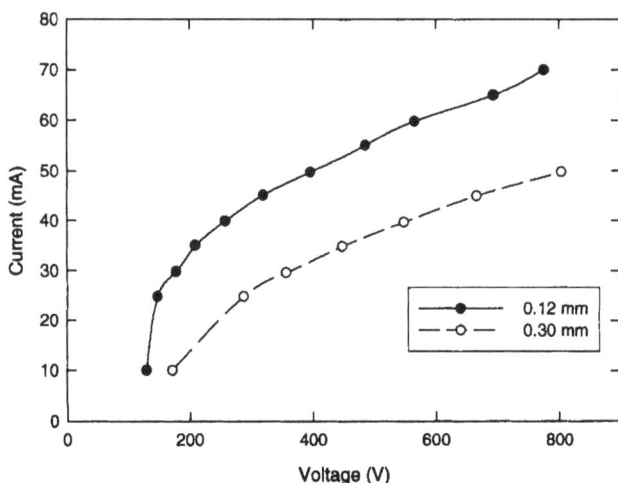

Figure 1.4 *Effect of changing anode-to-sample gap, from 0.12 to 0.30 mm, at constant pressure*

Figure 1.5 *Micrometer screw gauge adapted for measuring the anode-to-sample gap*[1]
(Reproduced with permission from John Wiley & Sons)

it should be kept constant for a given type of analysis. Variations in gap size will lead to variations in glow discharge conditions and thereby alter the analytical results.

The anode-to-sample gap can be measured with a specially adapted micrometer screw gauge (see Figure 1.5). Ideally, it should be capable of measuring the gap to within ± 10 μm. If no such device is available, one way of suspecting that the gap is getting too large is that it becomes more difficult to get flat craters, especially with typical source conditions.

5 Centring the Spectrometer

GDOES instruments containing a polychromator(s) built on the Paschen–Runge design, with a movable entrance (primary) slit, fixed exit slits and photomultiplier tubes as detectors, need occasional centring of the primary slit. This generally means the entrance slit of the spectrometer is moved to give maximum intensity on one specific line. The result is that all other lines on the polychromator will also be centred simultaneously. How well this works depends on how well the secondary slits were aligned during manufacture of the spectrometer. An example is shown in Figure 1.6.

Depending on the design of the spectrometer, this operation needs to be performed every day, occasionally or only once the polychromator has been opened. It is a good practice to use the same conditions (method) and the same sample and to note the changes in peak position and intensity. This will indicate if there is a problem in the spectrometer, such as change in temperature, or if the window needs cleaning.

Scanning monochromators in the Czerny–Turner design also require occasional alignment, because of the movement of the grating, typically using the zero-order reflection from the grating and argon reference lines. A useful argon reference line is at 415.859 nm.

Spectrometers using CCD detectors to cover wide ranges of the spectrum generally do not need this operation, as it can be integrated directly in the mathematical algorithms used for line identification. Nevertheless, the assignment between pixel number and wavelength should be checked from time to time and, if there are systematic errors, a new function should be applied.

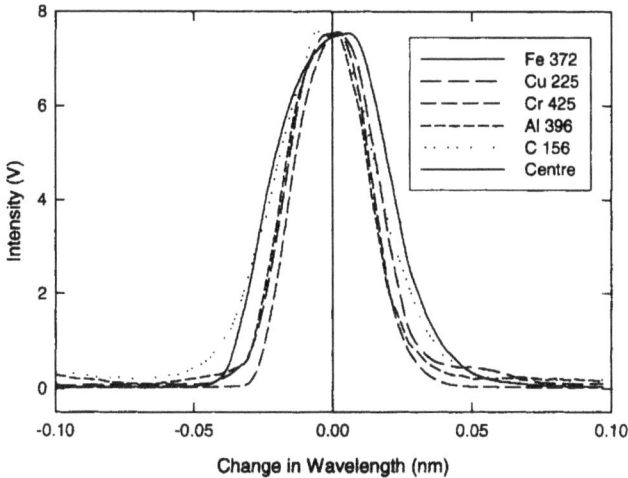

Figure 1.6 *Example line scan for Fe 372 from a steel sample used for aligning the poly-chromator, showing that other lines on the polychromator are also aligned simultaneously*

6 Window Cleanliness

Monitoring intensities over time for selected elemental lines covering a wide range of wavelengths allows checking the performance of the optical spectrometer. A continuous decrease of all intensities, being most severe at short wavelengths, would indicate increasing pollution of the optics, particularly the window between the glow discharge and the optical spectrometer. An example is shown in Figure 1.7. In this example, the intensities of the many lines on a polychromator were measured for two samples, Zn–Al alloy Canmet ZN3 and Steel NBS 1262, soon after cleaning the window and then 3 weeks later, after analysing a large number of other samples, including many paint coatings. The intensities after 3 weeks were then divided by the intensities with the clean window. After seeing these results it was decided, for this particular instrument, to clean the window every week. For other instruments analysing other materials, it is common to clean the window only once a month, or even less frequently. This decision will vary from laboratory to laboratory and is best decided from experience and from data over time.

Contamination of the window can occur from sputtered material. Grimm-type sources are designed to have a counter-flow of argon that travels across the window and is then directed towards the sample to minimise deposition on the window. Sometimes pollution comes from the oil pumps used in the source, and sometimes from the evaporation of low-temperature samples, *e.g.* tin or lead. Oil will tend to accumulate in places with the lowest temperature. Clearly, the source should not be at the lowest temperature in the pumping line. Additional pollution can come onto the back of the window from the outgassing of inappropriate sealants inside the spectrometer, and in vacuum spectrometers, from back streaming oil from vacuum pumps.

A common question is: how often to clean the window? As suggested above, there is no general answer, because of variations in source designs and varying workloads

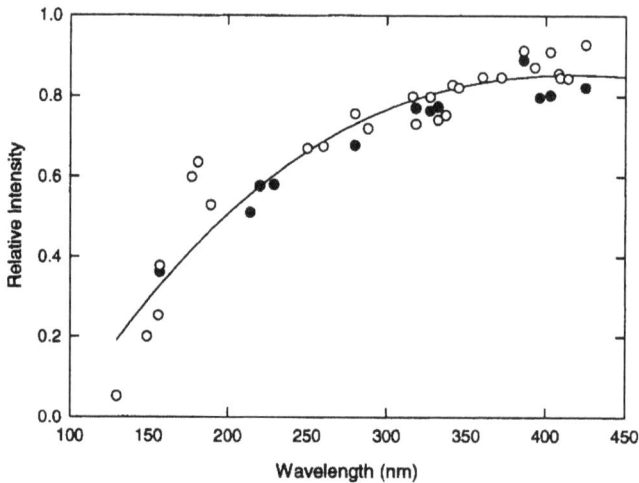

Figure 1.7 *Change in intensity of the many lines on a polychromator after 3 weeks:*
  • *Zn–Al alloy ZN3, ○ carbon steel NBS 1262*

on instruments. But the best approach is to monitor changes in intensities at various wavelengths over time, *e.g.* in run charts. Always use the same experimental conditions (method) and the same samples. If intensities are dropping, it is probably due to a dirty window. Cleaning the window should restore intensities to their earlier values. After a while, it will become clear how often cleaning is necessary.

Sometimes operators have noticed intensities drop quickly soon after cleaning the window and then more slowly. Under these circumstances, they prefer to clean less often, *i.e.* to accept lower intensities in exchange for long-term stability. Clearly, in these cases there is a source of pollution that only slightly covers the window. In an ideal world, it would be better to fix the source of pollution than accept lower intensities.

7 Designing a Standard Test

After these general considerations on checking instrument performance, many readers might now expect us to provide them with a plan for their routine checks, defining which parameters must be checked and how often. Considering, however, that there is really no such thing as a 'good' or 'bad' instrument, but just instruments that either meet specifications or do not, it is rather difficult to write such a general document.

International Standards Organisation (ISO) standard 10012 requires that all measurement tools, *i.e.* those used for serious measurement, must be checked to see whether they meet the requirements for the task or not.[2] These checks must be performed at regular intervals. The nature of the checks and their check interval must allow the user to assure that the measurement tool is adequate for use.

For a GDOES instrument, this means that all physical properties that may influence the analytical results, and we have identified many of them in this chapter, must be checked on a regular basis. To avoid spending more time checking the instrument than doing analysis, it is sensible to choose the intervals to be as long as possible. It is therefore suggested to monitor key parameters on a daily basis when the instrument is new. As the amount of data increases, you can derive new, hopefully longer, intervals based on the variations in parameters, their influence on the final analytical result and your analytical requirements (precision and accuracy). This ensures that you meet ISO standards, and that your analytical results are as good as you need them to be.

8 Shutdown

There are two approaches for shutting down an instrument for the night:

- Leave everything on, on the basis that being on promotes thermal stability and minimises interruptions and possible power spikes.
- Turn off as much as possible, on the basis of saving energy and prolonging life by minimising use.

Each has its advantages and disadvantages. Leaving everything on will mean:

- fast start up in the morning
- possibly a more stable instrument, but
- more prone to power outages overnight
- more pollution from the oil pump(s)

The disadvantages can be minimised by ensuring the instrument will restart correctly after a power failure, and by having a small argon leak into the pumps to reduce the back diffusion of oil. Note that not all instruments use oil pumps.

Turning most things off will mean:

- saving energy
- possibly extending the life of the instrument, but
- possible damage from surges during start up
- slower start up in the morning, especially waiting for the oil pump(s) to reach equilibrium temperature (typically 1 hour)

The disadvantages can be minimised by always starting in the same sequence and by having oil pumps turn on automatically with a timer about 1 hour before starting work.

Whichever approach is adopted, it is advisable to keep the source closed overnight, to reduce contamination of the source by air and moisture, which could then affect early analyses the next day. This can be done by placing a flat sample (*e.g.* a glass slide, a small metal sheet or the cooling block, if appropriate) on the source and then closing the valves to the pump(s) so that the source is left under vacuum but without argon and without being pumped. The pump(s) may then be turned off if you wish. This exact procedure may not be possible on all instruments but it should be possible to place some form of temporary cover on the source overnight. Check that the procedure used does not mean that a large quantity of argon is being consumed all night.

Nitrogen-purged spectrometers should be kept purged overnight; large vacuum spectrometers should be kept evacuated overnight. Should it be necessary to stop the nitrogen or vacuum pumps, perhaps for other maintenance in the laboratory, care should be taken when switching off the instrument to ensure the spectrometer is left in a suitable state. Refer to your user's manual. Later, when restoring the instrument, purging or evacuating the spectrometer to a level where intensities in the far UV are stable may take several hours. It may also take some hours for the spectrometer to reach a constant temperature, in particular when the operating temperature of the spectrometer is above the temperature of the laboratory environment.

For vacuum spectrometers, intermediate pressures, *i.e.* pressures between about 10 and 10 000 Pa (0.1 and 100 mbar), are dangerous for photomultipliers, because a glow discharge can be ignited around the tubes at the high-voltage supply connection. Normally, the electronics does not allow the high voltage to be switched on at these pressures. During pump-down, therefore, high voltages to the spectrometer should be switched off.

9 Start-up and Shutdown Sequences

The following steps are recommended for routine *start up*:

1. check gas supplies, *i.e.*, argon (and nitrogen)
2. check water supply for open systems or check level of cooling liquid for closed systems
3. run a common sample for several minutes
4. centre polychromator, if needed

The following steps are recommended for routine *shutdown*:

1. if possible place a light, flat sample or cooling block on the source, evacuate and close valves to pump(s), otherwise place a cover over source opening
2. if turning off source vacuum pump(s), turn off pumps
3. if necessary turn off argon
4. keep spectrometer either evacuated or with nitrogen

References

1. D.G. Jones in *Glow Discharge Optical Emission Spectrometry*, R. Payling, D.G. Jones and A. Bengtson (eds.), John Wiley & Sons, Chichester, 1997, 160–4.
2. Norme Internationale ISO 10012-1, Exigence d'assurance de la qualité des équipements de mesure, Partie 1, 1992 [French/English].

CHAPTER 2
Samples: Selection and Preparation

One of the attractions of GDOES is its ability to analyse solid samples directly with minimal sample preparation. But some amount of sample preparation is normally necessary, and proper sample selection and preparation can improve the accuracy and quality of an analysis. Sample selection and preparation are especially important when calibrating or drift correcting a GD spectrometer.

1 Sample Selection

Studies have shown that in geochemical and environmental investigations, for example,[1] the largest source of error in chemical analysis is often not due to instrumental performance, but to sample selection, or should we say 'miss-selection'. Errors in sample selection can range from choosing samples that do not fully represent the larger material, to mislabelling samples. If there is a large uncertainty in sampling then it is not an effective practice to deal only with improvements in the instrumental technique. Hence we need to consider measurement as just a part of the whole process.[2]

The magnitude and effect of sampling errors can be estimated by occasionally analysing additional samples, and by closely cooperating with the customer providing the samples. It is a good idea for analysts to check their analytical results directly with the customer's data to check for problems, or bias, in sampling.

GDOES requires very homogeneous samples! The typical inner diameter of the anode tube and therefore the diameter of the burn spot vary from 2 to 8 mm. The two extremes represent surface areas of 3 and 50 mm^2, respectively. The sputtering rate is typically some micrometres per minute; let us assume, for example, it is 5 μm min^{-1}. In bulk analysis mode, signals are integrated over a period of about 10 s. If the instrument uses a 4 mm anode, for example, a volume of about 0.25 mm^3 is used for the analysis; and, if we assume a density of 8 g cm^{-3}, comparable with steel, we use only about 2 mg of the sample. This 2 mg mass should represent the *entire* sample. We can immediately understand why GDOES requires homogeneous samples for analysis. Fortunately, the 2 mg of the sample also constitutes about 10^{18} atoms, so if

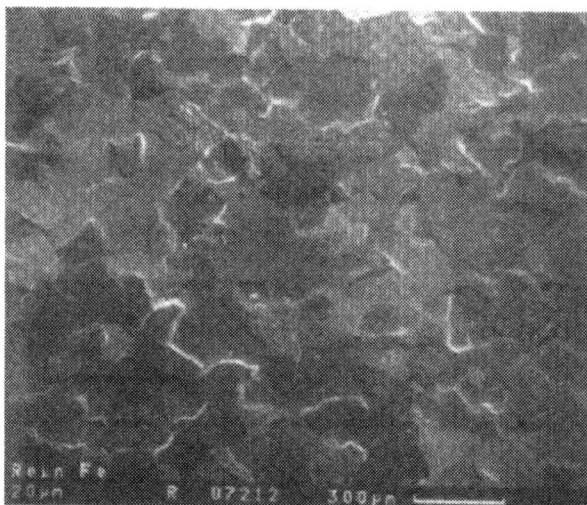

Figure 2.1   *The crater bottom on a steel reference sample with large grains, showing the effect
of differential sputtering on the different grains*[3]
(Reproduced with permission from J. Angeli, in *Glow Discharge Optical Emission
Spectrometry*, R. Payling, D.G. Jones and A. Bengtson (eds), John Wiley & Sons,
Chichester, 1997, 318–28)

this mass is truly representative of the complete sample, then the measurement has
an analytical merit.

The same reasoning can explain why grain structure may influence the quality
of the analytical results. Different grains can sputter at different rates, the so-called
differential sputtering. This can make the crater bottom look like a lightly etched
surface (see Figure 2.1). If the grains are large compared to the crater diameter, then
different areas on the sample could sputter at different rates, giving variable emission
signals for elements in different parts of the sample. Depending on the manufacturing
process, some trace elements may be enriched at grain boundaries. When grains are
large compared to the sputtered volume, the ratio of grain boundary to grain volume
may vary significantly from one analysis to the next, thereby influencing the quality
of the analytical results for trace elements.

The manufacturing process for producing homogeneous certified reference materials (CRMs) has improved significantly over the last 20 years. It is becoming apparent
in the industry that newer sets or classes of CRMs give analytical results with much
better precision than those obtainable with older CRMs of the same type. In addition,
manufacturers of CRMs have begun publishing on the certificates both the average
compositions of the CRMs and their estimated uncertainties. The published uncertainties are a valuable means for estimating the reliability of our own measurements.

Given the significant cost in purchasing CRMs, most analytical laboratories cannot
afford to keep buying new sets of CRMs all the time. Often, the analyst is forced to
use whatever materials are available. In this case, be aware that the final results will
never be more reliable than the materials used to calibrate or verify the instrument.

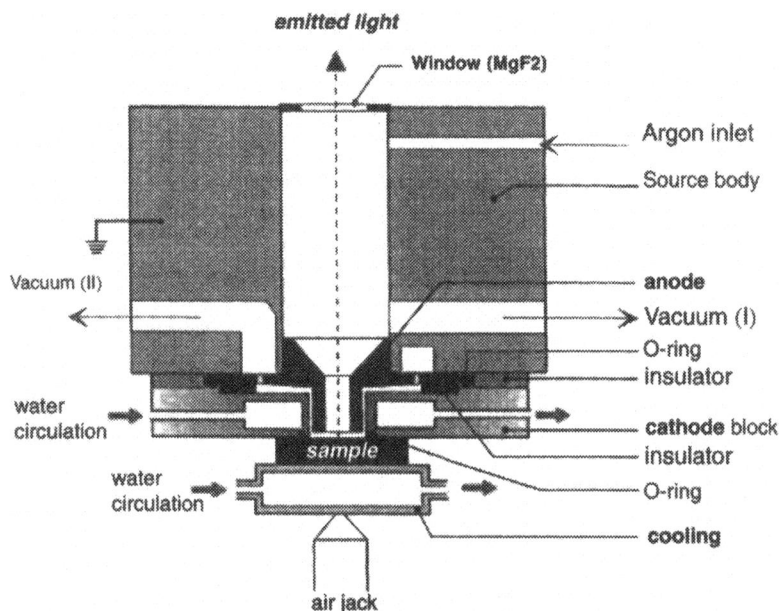

Figure 2.2 *Diagram of Grimm-type glow discharge source*[4]
(Reproduced with permission from M. Bouchacourt and F. Schwoehrer, in *Glow Discharge Optical Emission Spectrometry*, R. Payling, D.G. Jones and A. Bengtson (eds), John Wiley & Sons, Chichester, 1997, 54)

2 Preparation of Bulk Samples

Sample preparation is a much-discussed issue for GDOES. Supporters of GDOES will often say that it requires no sample preparation, while those supporting other techniques will sometimes say that sample preparation in GDOES is extremely tedious. Reality, as usual, falls somewhere between these two extremes. To understand the requirements for sample preparation and surface finish in GDOES, let us look at the commonly used Grimm-type GD source, and recall some of its physical properties (see Figure 2.2).

The glow discharge source is operated at low pressure. The discharge cavity is closed when the sample compresses a silicone O-ring forming an airtight seal. The properties of the discharge plasma depend very much on the quality of the carrier gas, usually argon. Small amounts of other gases, especially hydrogen and nitrogen, may significantly alter the analytical results. The source impedance, and hence the nature of the plasma, also changes with the anode-to-sample gap.

We can easily deduce that samples should be flat (at least in the region near the anode and in contact with the O-ring), and the vacuum seal obtained by the O-ring and the sample should be as tight as possible. Mirror polished sample surfaces are therefore the safest way to obtain reproducible analytical results. However, in many cases, even a roughly ground sample will give good results. To understand this apparent contradiction, we need to look at the problem in more detail.

Table 2.1 *Recommended preparation*
techniques for different materials

Material	Surface finish
Steel	Paper 180–600
Cast iron	Paper 180–600
Cr, Ni	Paper 180–600
Copper, brass	Machine
Aluminium	Machine, avoid SiO_2 or SiC paper

The performance of the GDOES source depends on many different parameters, such as the quality of the optics, stability of power supplies, cleanliness of the total vacuum system, condition of rotary vacuum pump oils, *etc.* All variations will accumulate to produce a deviation of the measurement from its expected value. Once the influence of surface finish is less than the influence of other variations, there is not much sense in just improving the surface polish.

In OES, the simplest way to represent spectral lines is by element and wavelength (in nm). Some spectral lines used for analysis react more strongly to perturbations in the plasma conditions than others. For example, Si 288.158 is strongly affected, while Ni 341.476 is barely affected. When the spectral lines for certain elements, more tolerant to plasma disturbances, are analysed, the influence of a rough surface will be fairly small or negligible.

As mentioned before, the sample is pressed onto the silicone O-ring and this in turn seals or closes the vacuum chamber. A rough sample surface, therefore, may allow air to leak into the vacuum chamber (GD source). Sharp edges that can pull the sample away from the O-ring and scratches on the sample surface have a bad influence on the quality of the analytical results.

Recommended methods for the preparation of some common metals are listed in Table 2.1. Use of dry abrasive papers to polish soft materials such as aluminium and brass should be avoided. When aluminium is polished with dry SiO_2 paper, for example, detection limits of only 0.1% Si in nearly pure aluminium have been observed.

Machining of Bulk Samples

Some materials are best prepared for calibration and bulk analysis by machining with a lathe or milling machine. These materials include aluminium alloys and copper alloys, such as brass. The sample is placed on a lathe and turned onto a cutting tool to remove sufficient amount of the metal surface to ensure a smooth, flat, fresh surface. This technique avoids contamination of the surface by polishing grits, and also avoids curvature of the surface often introduced by hand polishing. Care is needed to ensure that the cutting is not too fast as this may produce a series of rings on the surface that could allow air to ingress past the O-ring seal into the source. Cutting lubricants and oils should also be avoided.

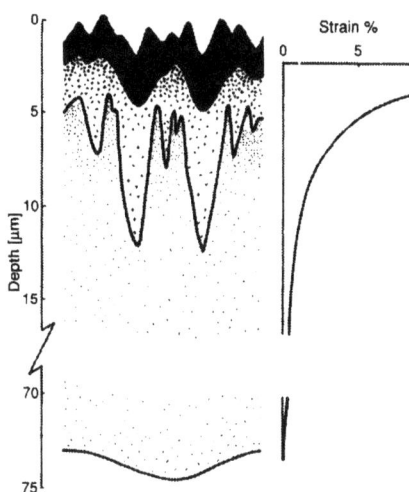

Figure 2.3 *Diagram showing the deformed layers on polycrystalline brass abraded with 220-grade SiC paper. Also shown is the approximate form of the strain gradient*[5] (Reproduced with permission from L.E. Samuels, *Metallographic Polishing by Mechanical Means*, 2nd edn, Pitman & Sons, Melbourne, 1971, 27–63, 70)

Polishing of Bulk Samples

When machining is not appropriate (*e.g.* hard materials, such as steels), bulk samples should be freshly polished before analysis. It is not absolutely necessary to obtain a mirror finish; many operators finish with 200-grade papers. But it is essential that the polishing or grinding be as flat as possible and without any deep scratches or pits. This reduces potential contamination of the sample surface with embedded polishing grit residue. Flatness is important to ensure that the anode-to-sample distance is constant across the sample. Scratches and pits can allow air to leak into the plasma past the O-ring seal. Embedded grit can add unwanted intensities and so bias the analysis; for example, polishing steel with SiC grit can add small amounts of Si and C during the analysis.

Wet polishing, followed by forced air-drying, is normally the preferred method compared to dry polishing. Wet polishing causes less heating of the sample surface (*e.g.* better C results in steel and iron), reduces diffusion and is less likely to clog the abrasive paper and leave grit on the surface. This generally results in fewer large and deep scratches because polishing debris, *e.g.* inclusion particles, is more easily removed than with dry polishing.

It is advisable to polish the samples on the same day for the analysis in order to avoid oxidation. This is especially true for rapidly oxidising materials such as zinc alloys, and materials with inert oxides such as aluminium alloys. Wherever possible, it is also advisable to polish calibration samples and analysis samples exactly the same way. This ensures the pre-burn is equally effective on the two sets of samples.

Polishing damages the surface of the sample, that is, it alters its structure and chemistry (see Figure 2.3). As a rule of thumb, the depth of damage is equal to the size of the polishing grit. Hence 5 μm grit will damage the surface to a depth of about 5 μm. If the sputtering rate is about 5 μm min^{-1}, then it will typically take a minute to sputter through this damaged region into the normal bulk material.

When polishing with 220-grade SiC paper, for example, about 20% of the contact points are involved in cutting chips off the surface, the rest are simply ploughing grooves in the surface.[5] The major work is therefore in damaging the surface rather than removing it. When viewed in cross-section, the damage on a metal surface appears to have three zones: surface scratches, gross deformation below these, and deep deformation. For brass abraded with 220-grade SiC paper lubricated with water, these regions are about 2 μm, 8 μm and 80 μm deep, respectively, with the depth of damage reducing with the age of the paper. The outer damaged layers contain very fine grains and an extremely high density of defects. How this restructuring of the surface relates to chemistry affecting GDOES intensities has not been studied, but the less the surface resembles the bulk material, the more likely will it negatively affect GDOES results.

Fresh dry papers initially abrade similar to wet papers but very quickly become clogged and begin causing deep rough grooves, the finer the grade the faster the clogging. With dry abrasion on steel samples, surface temperatures can exceed 650 °C, sufficient to cause thermal diffusion and re-alloying. Even a modest amount of water will prevent such high temperatures. Embedding of abrasive particles in the sample surface is much more severe in dry polishing.

Typically, samples are given a coarse first abrasion/grind to remove previous GD craters. This usually causes serious damage to the remaining sample surface. This surface is then abraded with a finer paper, typically 220-grade, to remove any visible deep scratches. But it is better to continue polishing with the finer paper beyond this stage to remove more of the deeper damaged zones. Often, operators stop at this stage, though some continue polishing with even finer grades, *e.g.* 400-grade. A high level of polish is not normally necessary for chemical analysis but can greatly assist in measuring crater shapes and depths with a profilometer, because of the reduction in surface roughness.

3   Surface Samples

Surface finish for coating analysis obviously depends very much on the purpose of the analysis. In most cases no surface preparation is required, or even desirable, as any surface treatment will alter the surface chemistry and may remove the very surface features required for analysis.

Unless the oil on the surface is the subject of the analysis, highly oiled samples should be degreased before analysis to obtain the best depth profiles of the surface and coatings. Besides being unnecessary for the analysis, oil contains large amounts of C, one of the slowest sputtering elements. C also tends to be knocked into the surface by argon bombardment. Hence, superficial oil layers can be very slow to remove during sputtering and can last for a longer time during depth profiles, see Figure 2.4, where the RF-GDOES depth profiles for an oiled surface, before and after degreasing with acetone, are superimposed. Note that such layers can cause instabilities in the plasma for DC-GDOES.

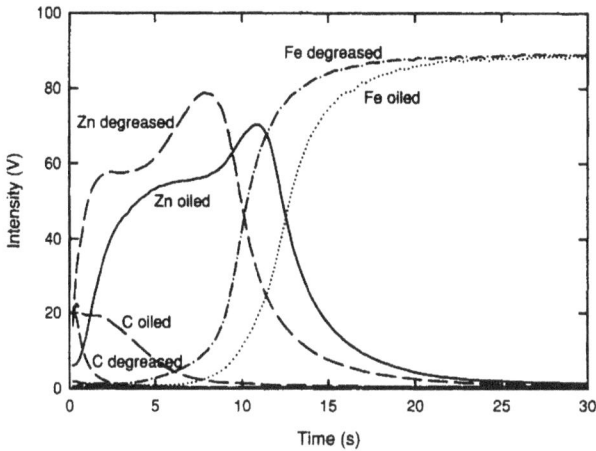

Figure 2.4 *Oiled, thin galvanised coating on steel, by RF-GDOES, as received and after degreasing in acetone*

Thick grease films are non-conductive and, usually, non-homogeneous in thickness. Their presence on the sample surface significantly decreases the depth resolution. This effect, though exists for RF and DC discharges, is more important when DC is used. Thick grease layers, often present in industrial samples, may completely stop a DC discharge from being ignited and therefore make the analysis impossible.

Oil, grease films, fingerprints, *etc.* can perturb the analysis of the extreme surface for several different reasons. We have mentioned reduced sputtering rates and reduced depth resolution. But such contaminants also contain the element H. The effect of H on the emission yields of different spectral lines can be very strong, and though this effect can, in principle, be corrected in software it is better to avoid it from the beginning. Samples for surface analysis should, therefore, be handled carefully to avoid contamination of the surface to be analysed.

Industrial coatings of several micrometres thickness may be wiped first with a soft tissue, eliminating most of the grease. The sample may then be rinsed in alcohol, acetone, petroleum spirits or detergent liquids. Any residuals should be rinsed away with alcohol or acetone or a similar quick drying solvent. An ultrasonic bath is very efficient for cleaning; however, care should be taken not to damage the coating.

Care should also be taken in choosing solvents. Not only can they be hazardous to humans, but they can also leave residues on the surface. Some low-quality acetones, for example, contain oil. Before using a solvent for the first time, it is advisable to check material safety data sheets and then check the level of contamination from the solvent on a clean highly polished sample using a GDOES depth profile.

4 Small Samples

Special holders are available for mounting small samples onto the GD source. An example is shown in Figure 2.5. The sample no longer seals the source with the usual O-ring to sample seal, but rather the sample holder itself is used to seal the source.

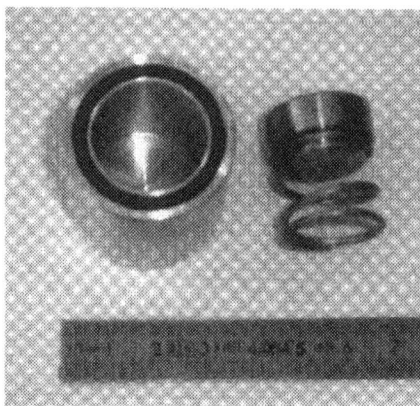

Figure 2.5   *Small sample holder. The sample is mounted on a spring-loaded brass disc (scale in cm)*

To ensure that the plasma is fully obstructed, the sample width must be greater than the outside diameter of the anode. Typically, this means the sample must be at least 1 mm wider than the inside diameter of the anode, which is the diameter normally quoted. Therefore, for a commonly used 4 mm anode, the sample must be at least 5 mm wide, while for a 2 mm or 2.5 mm anode it can be as small as 3 mm wide.

Care is obviously needed to ensure that the small sample is located exactly on the central axis of the anode. Thin samples should be mounted on a flat, smooth metal surface using a conductive tape (such as a double-sided copper tape or carbon dots) to prevent them from bending towards the anode under vacuum and causing a direct short circuit to the anode.

Large, irregularly shaped or rough samples can be difficult to mount on the GD source in the normal way because they will not seal properly on the O-ring. It can be worthwhile cutting such samples so that they can be mounted on the small sample holder. Similarly, porous samples can also be dealt with.

5   Powders

Various materials can be analysed by GDOES as powders after some preparation. Such materials include metal powders and crushed rock.

Powders must initially be ground to a very fine state and then mixed thoroughly with a binder before compacting under high pressure into a solid pellet. The goal is to create a compact, non-porous pellet with high homogeneity. To achieve this, the grinding and mixing are critical. The normal ratio of analyte powder to binder is 1:10, though ratios of 1:4 to 1:19 have been used successively.[6] The dilution of the analyte powder reduces element intensities and hence detection limits. The most commonly used binding material is copper powder (*e.g.* MERCK No. 2703), though silver and gold are also used, especially if Cu is one of the elements to be analysed.

A high compacting pressure is normally required. Depending on the material, up to 800 MPa may be necessary, held for 2–3 min to allow time for plastic flow and release slowly to prevent cracking.

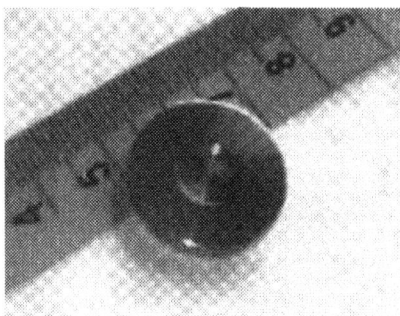

Figure 2.6 *Special powder holder. The powder is pressed into the small cavity in the copper disc (scale in cm)*

The pellet should be about 2–3 mm thick so that it is rigid and vacuum-tight. A typical pellet, 25 mm in diameter and 2.5 mm thick, would require about 6 g of powder (0.5 g of analyte powder). The quantity of powder required could be reduced by recalling that the face of the pellet to be analysed needs only be slightly larger in diameter than the anode being used. One solution is to place the powder into a special holder, similar to the one shown in Figure 2.6, before pressing. The seal is then made by the copper holder rather than by the pressed powder. The cavity shown has a diameter of 7.5 mm and a depth of 1.5 mm, requiring only about 20 mg of analyte powder.

Those intending to analyse compressed powders using GDOES should be aware that it is not a straightforward exercise. Reacting strongly with any air introduced into the plasma chamber is one of the analytical properties of the plasma. So, pellets should definitely contain as little residual gas as possible.

6 Rods, Tubes and Wires

Some curved samples can be analysed by GDOES, while others are quite difficult to do so. It depends mainly on the radius of curvature. Curved samples might include rods, tubes, wires, balls and lenses. Samples with a large radius can be mounted directly on the source, since the anode-to-sample distance will be constant and the sample will not touch the anode (see Figure 2.7). Some tubes can be pressed flat and then analysed like flat samples. Small-diameter samples can be laid side by side and pressed into a soft metal substrate such as lead.

For special applications, such as coatings on tubes, the front face of the anode can be sculpted to match the curvature of the sample; this ensures a constant anode-to-sample distance.[7] Although it is not a common practice, wires and small-diameter rods can be mounted on the source as 'pin' samples, where the sample sits along the central axis of the anode, by inserting the sample perpendicularly into a hole at the centre of a metal support.[8]

Although the analyses of wires, rods and some odd-shaped samples are possible with GDOES, and have been described in the literature, they are far from being a standard application. In fact, each new shape may require a new and challenging study, before the sample can be mounted on the GD source, and the precision of the

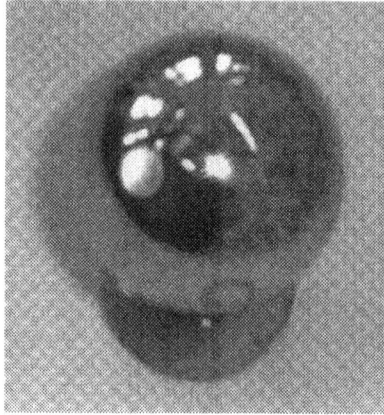

Figure 2.7   *A curved sample (artificial hip joint, 20 mm diameter) analysed successfully on a standard GD source, as evidenced by the many craters*

obtained results assured. This will entail an additional analytical cost in the development of the new method and the hardware needed. It should also be pointed out that quantitative analysis of powders, rods and other samples that have a special shape requires extra care. Significant errors can occur when a calibration is performed with flat bulk reference materials and this calibrated method is then used for calculating the elemental composition of a powder or a non-flat sample.

References

1. M.H. Ramsey, *J. Anal. Atom. Spectrom.*, 1998, **13**, 99.
2. F. Adams, A. Adriaens and A. Bogaerts, *Anal. Chim. Acta*, 2002, **456**, 63.
3. J. Angeli, in *Glow Discharge Optical Emission Spectrometry*, R. Payling, D.G. Jones and A. Bengtson (eds), John Wiley & Sons, Chichester, 1997, 318–28.
4. M. Bouchacourt and F. Schwoehrer, in *Glow Discharge Optical Emission Spectrometry*, R. Payling, D.G. Jones and A. Bengtson (eds), John Wiley & Sons, Chichester, 1997, 54.
5. L.E. Samuels, *Metallographic Polishing by Mechanical Means*, 2nd edn, Pitman & Sons, Melbourne, 1971, 27–63, 70.
6. W. Fischer, in *Glow Discharge Optical Emission Spectrometry*, R. Payling, D.G. Jones and A. Bengtson (eds), John Wiley & Sons, Chichester, 1997, 149.
7. F. Schwoehrer, in *Glow Discharge Optical Emission Spectrometry*, R. Payling, D.G. Jones and A. Bengtson (eds), John Wiley & Sons, Chichester, 1997, 155.
8. S. Hagstrom, private communication, 2001.

CHAPTER 3

Optimisation of Source and Analytical Parameters

The optimisation of source and analytical parameters is an important part of performing reliable and precise GDOES analysis. Even for standard applications, such as the bulk analysis of low-alloy steel or the depth profiling of galvanised steel sheet, where it is possible to list the optimal source conditions for different instruments, the operator should try to understand the influence of the different parameters on the analytical results. Very often, compromises have to be made between the quality of an analysis and the time (cost) of the analysis. Without this understanding, when selecting or varying source or analysis parameters, the operator risks either the quality of the results or the loss of time by being overcautious.

In most cases, GDOES analysis is performed in three steps:

- flush
- pre-burn (pre-integration)
- integration of results

For obvious reasons, the pre-burn is generally skipped for surface and coating analysis.

1 Flush Time

Each time the sample is removed from the source, the source is open to air. The discharge cavity will therefore contain some air every time the sample is changed or moved to a new position, even if the source is flushed with argon during the positioning process, though a constant argon stream should reduce the ingress of air. To obtain a pure argon atmosphere inside the source, after placing the sample, the closed source cavity is first evacuated and then flushed with argon. An effect of flushing the cavity prior to analysis is to reduce the time for the plasma to become stable during the analysis.

Different flushing procedures are available for different instruments:

- simple argon flush
- alternating evacuating and flushing
- pulsed argon flushes

They are all effective, but depending on the types of vacuum pumps and the design of the discharge cavity and connecting vacuum tubes, one or the other may be more efficient. When oil-based rotary pumps are used, a long evacuation of the discharge cavity down to residual vacuum should be avoided, as it will promote pollution of the source by back streaming of hydrocarbons from the pumps.

Two different types of pollution occur when opening and closing the discharge cavity:

- residual gases (including N_2, O_2, CO_2, H_2O from air)
- deposition of H_2O vapour onto the cavity walls

Gases in the argon atmosphere are removed most efficiently with a strong and prolonged flow of pure argon towards the vacuum pumps. Adsorbed vapours on the cavity wall are more difficult to remove as collisions of argon atoms with the walls of the vacuum chamber generally have too low energy to remove adsorbed species easily. The most efficient means is increased wall temperatures; hence GD cells are often kept warm.

It is obviously difficult to define an optimal flush time for all applications and all needs. However, in our experience a flush time of 10 s is sufficient for most bulk analysis. Only when it is necessary to analyse at trace levels (below 100 ppm) elements such as N, O, H and C, which could be present as residual contaminants of argon, would it be necessary to increase the flush time to 1 or 2 min.

For surface analysis, as no pre-burn is used, the flush time should generally be longer, perhaps 30–40 s for typical cases and 1–3 min for the analysis of low levels of contaminating elements or when the extreme surface is to be analysed.

2 Pre-burn (Pre-integration)

Pre-burning the sample, also called pre-integration, simply means igniting the discharge, and waiting before integrating the spectral intensities. The purpose of pre-burning is similar to flushing the cavity, except that pre-burning removes oils, oxide films and other contaminants from the sample surface. Igniting the discharge also helps remove vapours, such as water, from the discharge cavity by destroying their molecular structure. The result is an improved relative standard deviation (RSD) for calibration and bulk analysis measurements.

Typical pre-burn times are between 30 s and 90 s. This corresponds to a sputtered depth of about 5–10 μm. It has been the experience of many people that pre-burn times can be reduced for well-polished samples compared to very rough ones. In a few cases, fairly long pre-burn times are required. Examples are grey cast iron, see Chapter 8, and the detection of C, N and O at low contents (< 10 ppm for C; < 100 ppm for N, O), where pre-burn times of more than 3 min are often recommended.

The optimal pre-burn time for a given application can be established most readily by simply running a depth profile on a typical sample to be analysed. An example is shown in Figure 3.1 for a bulk chill-cast iron sample. A quick glance at Figure 3.1a suggests that a pre-burn time of about 30 s might be sufficient, but a closer look, after

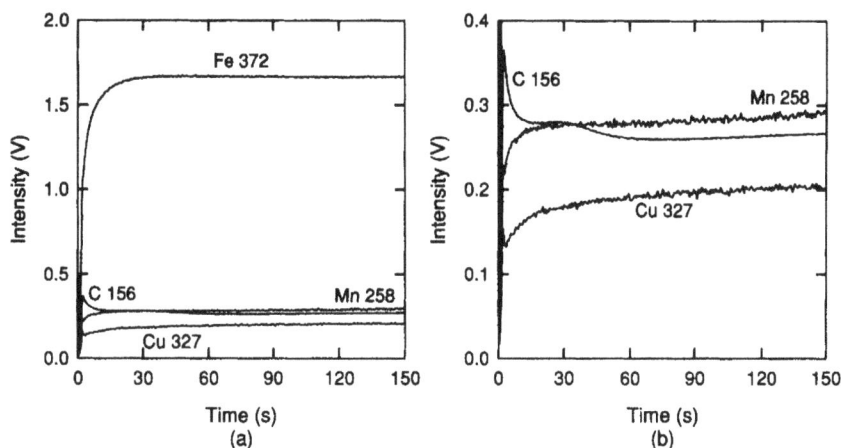

Figure 3.1 *Qualitative depth profile of a chill-cast iron: (a) showing major element Fe and minor elements C, Mn and Cu and (b) expanded intensity scale*

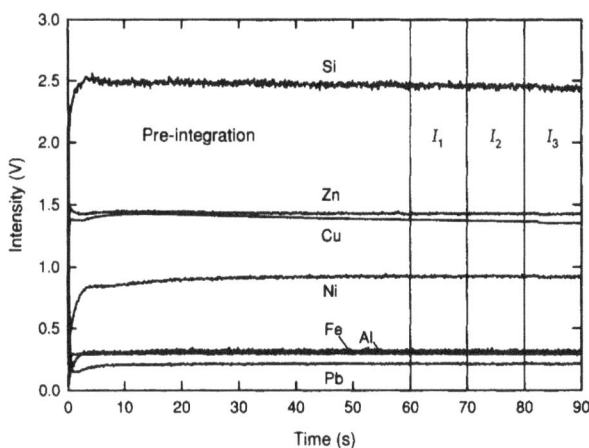

Figure 3.2 *Qualitative depth profile of brass CTIF L6, showing major and minor elements, indicating the time regions that would be used for pre-integration (60 s) and three integrations (I_1, I_2, I_3) each of 10 s*

expanding the C intensity scale, suggests about 80 s is needed. The sputtering rate was about 5 μm min^{-1}, so 80 s corresponds to about 7 μm.

Let us consider a second example in more detail: brass. We will use this material extensively throughout succeeding chapters, to illustrate the procedures for optimisation, and for calibration and analysis. The qualitative depth profile for certified brass sample CTIF L6 is shown in Figure 3.2.

Examination of Figure 3.2 suggests that a pre-integration time of about 50–60 s would be sufficient to establish stable signals. For the conditions chosen: 4 mm anode, 30 W RF, 800 Pa, this corresponds to a depth of about 9 μm.

Table 3.1   *Determination of optimum pre-integration time from changes in intensity with sputtering time of Ni 341 in brass CTIF L6*

Time (s)	1	2	3	4	...	8	Mean	SD
0–10	1.422	1.492	1.487	1.511		1.543	1.480	0.048
10–20	1.455	1.567	1.547	1.536		1.634	1.524	0.061
20–30	1.477	1.576	1.585	1.539		1.630	1.534	0.060
30–40	1.479	1.588	1.595	1.561		1.620	1.542	0.057
40–50	1.490	1.607	1.597	1.572		1.611	1.553	0.051
50–60	1.505	1.612	1.612	1.555		1.592	1.553	0.048
60–70	1.515	1.611	1.610	1.558		1.612	1.561	0.044
70–80	1.508	1.636	1.615	1.566		1.590	1.566	0.046
80–90	1.526	1.632	1.617	1.590		1.597	1.572	0.044
90–100	1.528	1.636	1.640	1.602		1.600	1.581	0.046
100–110	1.531	1.642	1.644	1.598		1.613	1.588	0.044
110–120	1.540	1.645	1.654	1.597		1.609	1.591	0.043

Figure 3.3   *Brass sample CTIF L6 showing the many craters used in the various tests in this chapter. It is a good practice to avoid using the centre of solid samples; the composition there often varies considerably from certified values due to segregation during solidification*

To check the optimum pre-integration time further, consider the following simple experiment:

Repeat time-resolved data acquisitions of about 180 s each, eight or more times, changing the analysis position after each run

The brass sample CTIF L6 is shown in Figure 3.3 and the results are shown in Table 3.1.

When the standard deviation (SD) either does not improve any further or is sufficiently good for one's needs, the pre-burn time is long enough and there is no need to wait any longer. The data in Table 3.1 confirm that 60 s is suitable for brass samples such as CTIF L6, using the conditions of the method.

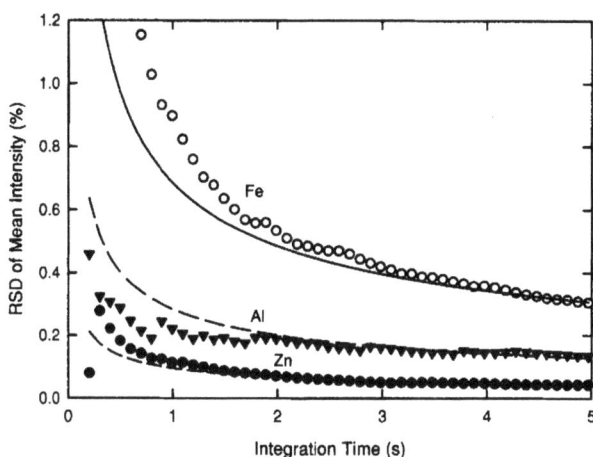

Figure 3.4 *Variation of uncertainty in mean intensity as a function of integration time for elements in brass CTIF L6. The solid lines are proportional to the square root of the integration time*

3 Integration (Averaging) Time

Whether it is for bulk analysis or for depth profiling, intensities are recorded for a finite amount of time. Historically, this time is called the 'integration time', because in early spectrometers the signals were recorded onto a photographic film and the recorded intensity varied with the time of exposure. Today, the signals are counted over a certain time and then divided by this time to give an average signal. Despite this, the measuring time is still commonly called integration time.

For bulk analysis, integration (averaging) times typically vary from 5 s to 30 s, and for depth profiling they are typically about 0.1–1 s but can be as short as 1 ms. Integration time is illustrated in Figure 3.2 for brass CTIF L6. Note in the figure that bulk analysis is just depth profiling with the data presented as averages.

If we have a constant signal with a normally distributed (white) noise, then increasing the integration time will not change the mean but will reduce the observed SD. If the signal varies during this time, increasing the integration time will restrict the speed at which the mean can follow the change in signal; this is important in depth profiling where signals can vary rapidly in going through different layers in the sample. If the integration time is too long, then the depth resolution will be adversely affected.

There are electronic time constants in the signal acquisition circuits that determine the shortest possible integration time and this means that even with this shortest acquisition time some measured reduction of the 'real' noise has occurred. For a constant signal with random (white) noise, the observed SD of the noise will decrease approximately as the square root of the integration time (see Figure 3.4). Hence an integration time of 10 s compared to an integration time of 0.1 s will reduce the observed noise to about $1/\sqrt{100} = 1/10$.

Table 3.2 *Demonstration of the effect of integration time on intensity for elements in brass CTIF L6, after 60 s pre-integration and total integration time of 30 s*

Signal	Integration (s)	Repetitions	Mean	SD	SD_{Mean}
Al	6	5	0.314 96	0.000 75	0.000 34
	0.1	300	0.314 96	0.002 52	0.000 15
Fe	6	5	0.362 36	0.000 98	0.000 44
	0.1	300	0.362 36	0.007 23	0.000 42
P	6	5	0.014 45	0.000 13	0.000 059
	0.1	300	0.014 45	0.001 20	0.000 069
Pb	6	5	0.267 30	0.001 9	0.000 85
	0.1	300	0.267 30	0.004 0	0.000 23
Zn	6	5	1.617 79	0.005 6	0.002 49
	0.1	300	1.617 79	0.006 4	0.000 37

But it is important to understand, in recording a good mean intensity, that it is the total integration time used to estimate the mean that matters. Repeating a measurement many times with a short integration time, though it gives an apparently larger standard deviation, can be better than a few measurements with a long integration time, provided the total integration time is the same. This is because the many short measurements give a better estimate of the standard deviation, as more information on the signal behaviour is available. In addition, averaging many short integration intervals allows for the detection and elimination of spikes.

The effects of variable integration time on the mean and SD are illustrated in Table 3.2, for brass CTIF L6. For all elements shown, no spikes were detected so the means were the same for 6 s and 0.1 s, since the total integration time was the same, 30 s. For Al, Fe and P, the SD of the measurements was much lower with 6 s rather than 0.1 s, as expected, but SD of the mean, SD_{Mean}, was similar, also as expected. The definition and importance of SD_{Mean} will be explained later. For such elements the total integration time could be extended well beyond 30 s. But for two elements, Zn and Pb, the intensities increased slowly during the 30 s total integration. For these elements, the SD_{Mean} using 6 s integration was much worse than that with 0.1 s, because of the contribution of the slope to the SD of the measurements. Because of this slope for these important elements, the total integration time for all elements was limited to 30 s during the subsequent calibration.

Unbiased Estimates

Figure 3.5 shows five measurements of intensity, I. If we knew the true mean μ (in this computer-generated example it is 1.000), we could estimate the variance of these measurements from:

$$\sigma^2 = \frac{\sum (I - \mu)^2}{n} \tag{3.1}$$

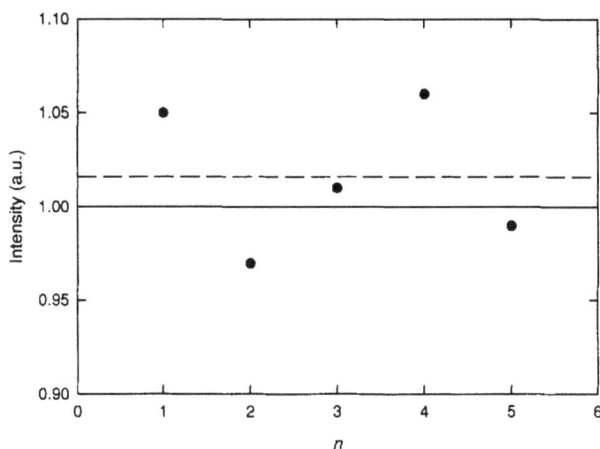

Figure 3.5 *Estimating the mean for test data whose true mean is 1*

That is, we would divide the sum of the squares by n, the number of independent values. But in spectroscopic measurements, we generally do not know the true mean. We can only estimate the true mean by averaging the measured data, $\bar{I}$, here obtaining 1.016, shown by the dashed horizontal line in Figure 3.5. We could then try to estimate the variance of the data about this dashed line by using Equation 3.1, obtaining 0.034. But since our estimate of the mean has already minimised the deviation of the points about the mean, our estimated variance will be biased low. Instead, to obtain a non-biased estimate, we use:

$$s^2 = \frac{\sum (I - \bar{I})^2}{n - p} = \frac{\sum (I - \bar{I})^2}{\nu} \tag{3.2}$$

where p is the number of fit parameters, *i.e.* the number of ways in which we can adjust the mean to fit the data, and ν is the number of degrees of freedom. In our example, $p = 1$ and we obtain $s = 0.038$, slightly larger than the biased value. A consequence of this method of estimating the variance is that n must be greater than ν; otherwise the variance cannot be estimated.

As we increase the number of measurements, we have less 'freedom' to shift the line because the shift will favour some points but will prove more and more disadvantageous for others, so that the estimated mean will approach the true mean and the estimated variance will approach the true variance.

This way of calculating variance is strictly valid only for white noise. Allan has proposed a different form of variance, often called the Allan variance or paired variance, which is more suited to time-dependent measurements with different types of noise[1,2]:

$$s^2 = \frac{\sum_{i=1}^{n-1} (I_{i+1} - I_i)^2}{2(n-1)} \tag{3.3}$$

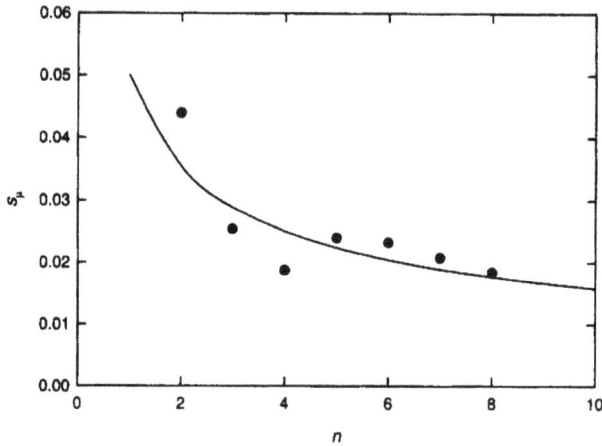

Figure 3.6 *SD of mean s_μ versus number of measurements n:* • *Zn 335 in brass CTIF L6*

where the squared term in the sum is the difference in intensity between two successive measurements. For random noise, Equations 3.3 and 3.2 will give the same value. Though the Allan variance is not used often in spectrochemistry, it is a useful way of checking the validity of the normal variance.

Repeating Analysis

When OES instruments are calibrated, the intensity measurement is usually repeated a few times and the average intensity is then used in the regression procedure. It is the 'reliability' of this average value (mean) that determines the confidence we can have in this data point.

Many analysts mistakenly think they should use the variance (or its square root, the standard deviation) of the measured intensities, i.e. the variance of the population, to estimate the uncertainty in the mean. This variance, s_p^2, is given by Equation 3.2. Assuming that we are dealing with intensity measurements having a normal distribution, carrying out more measurements will likely give us a more and more accurate measure of the population variance and hence the width of the distribution of intensities, but these extra measurements will not reduce this population variance nor appear to benefit the analysis.

For estimating the uncertainty of the mean and later for estimating regression parameters and their covariance, the variance of the mean should instead be used. Previously, we have used the symbol SD_{Mean} for this term but will now use s_μ. The estimated variance of the mean of n independent intensity measurements is given by:

$$s_\mu^2(\bar{I}) = \frac{1}{n(n-1)} \sum (I_i - \bar{I})^2 = \frac{s_p^2(\bar{I})}{n} \tag{3.4}$$

Equation 3.4 is illustrated in Figure 3.6 as a solid line, compared with intensity measurements for Zn from eight craters on brass CTIF L6. The fluctuations about the

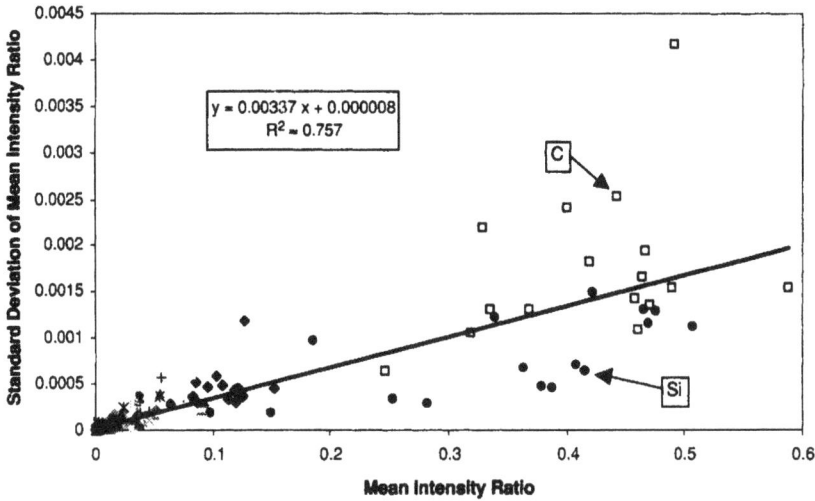

Figure 3.7 *Standard deviation of the mean intensity of 15 elements ratioed to iron, in cast iron, versus the mean intensity ratio*[3]
(Reproduced with permission from M. Winchester and J.K. Miller, J. Anal. Atom. Spectrom., 2001, 16, 22)

expected trend show the normal variation due to each added measurement. Despite such normal variations, the equation clearly shows the value in repeating intensity measurements. More measurements will tend to decrease the variance of the mean, and therefore increase the likelihood that the estimated value $\bar{I}$ is close to the true mean.

But here, a compromise between improved precision and longer time of analysis has to be found. For critical work, it is wise to repeat the measurement of each calibration sample five times.

Behaviour of s_μ

As part of their calibration for cast iron, Winchester and Miller plotted the standard deviation of the intensities for 15 elements as a function of the intensities. The values were ratioed to Fe. Their results are shown in Figure 3.7, with the regression line slightly modified to show a non-zero intercept. Though there are variations from element to element, as expected, because of variations in sensitivity from line to line (including different photomultiplier voltages), the trend is for the SD of the mean to increase almost linearly with intensity, *i.e.*:

$$s_\mu \approx s_0 + s_1\bar{I} \tag{3.5}$$

The RSDs are shown with log scales in Figure 3.8. These appear to decrease only slowly with increasing intensity except at low intensities, near the detection limit. Equation 3.5 is shown as the solid line.

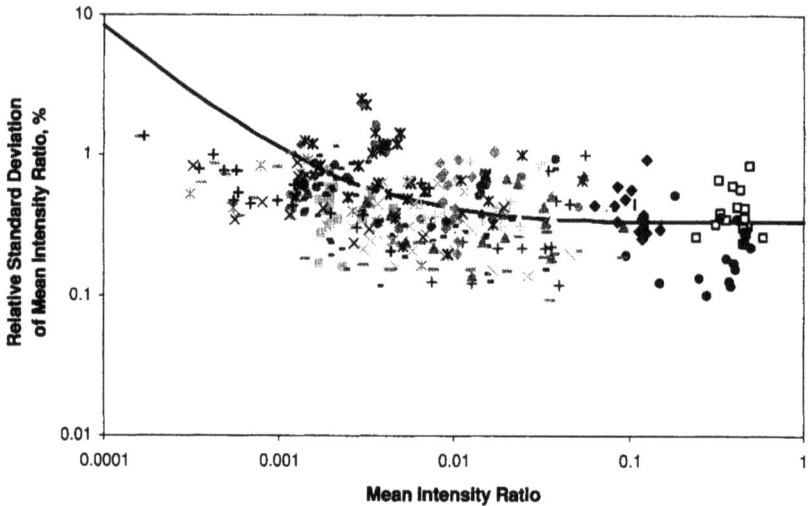

Figure 3.8   *Relative SD of the mean intensity of 15 elements ratioed to iron in cast iron versus the mean intensity ratio, from Figure 3.7, shown here with log scales*

Optimum Total Integration Time

One way to determine the optimum total integration time is by plotting the square of the SD of the mean, *i.e.* the variance of the mean, as a function of integration time. To illustrate this procedure, record the intensities repeatedly after pre-integration from a homogeneous sample using a short integration time (*e.g.* 0.1 s) for a much longer time then would normally be used for integration (*e.g.* 90 s). This could be done as part of a depth profile. Determine the variance of the mean for the full period. Then take only the first half and again determine the SD of the mean, and halve again and again, until the integration is too short to determine the SD reliably (*e.g.* <1 s). Now, plot the values for the variance of the mean as a function of the integration time. An example is shown in Figure 3.9 for brass CTIF L6.

Initially, as the integration time is increased, the variance of the mean will tend to decrease, as shown, but once the integration becomes too long, it may start to increase again, as shown for Zn. The variance of the mean for Zn increases because the intensities start to drift from their initial, stable value. The optimum total integration time is when the variance of the mean has just dropped close to its lowest value and well before it starts to increase again.

There are various reasons why the intensities will tend to change with time, even for homogeneous samples:

1. the sample surface is heated by interaction with the plasma and not all of this heat is carried away by the sample cooling circuit
2. material is redeposited on the edges of the crater, thereby changing the sample-to-anode gap; this, in turn, affects the efficiency of argon pumping at the sides of the crater, and changes the source impedance

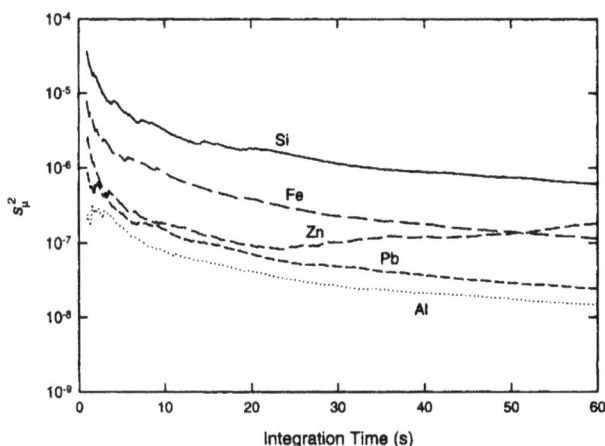

Figure 3.9 *Variance of the mean as a function of integration time, for brass CTIF L6, most elements showing the initial decrease expected for white noise but Zn 335 then shows a slow increase caused by changes in intensity with depth*

3. an increasing number of electrons can originate from the sides of the crater—making the sample appear more and more like a 'hollow cathode'

4 Precision: Internal and External

If we repeat a measurement several times during a single burn to determine repeatability, then this is called internal precision. If we repeat a measurement with several burns on a single sample, then the variation in the intensities is called external precision. They provide estimates of two different things. Internal precision estimates the noise in the signal and the short-term stability of the source and instrument. External precision includes variations due to noise and short-term stability, also it includes the variability in the sample (inhomogeneities in composition and surfacing) and in the mounting of the sample on the source (variations in anode to sample distance). External precision therefore tends to be greater than internal precision.

They are both expressed in terms of SDS, s_I and s_E, and their variances, s_I^2 and s_E^2, respectively. Typically, internal precisions for major elements in metals, or for minor elements ratioed to the major element in metals, are less than 1%, while external precisions for high quality, homogeneous metals that are carefully polished are typically less than 2%.

In calibration and bulk analysis, it is becoming a common practice to record mean intensities by recording results from several burns with several measurements (repetitions) in each burn, *i.e.* to combine internal and external precisions to determine the mean intensities and their SDs. This is certainly an interesting approach as more information is available to estimate the uncertainty of the mean and thus to improve the quality of the analysis.

Though the principle for estimating the uncertainty of a mean of n measurements, through uncertainty propagation, is fairly straightforward, in reality it can be confusing at times. In the following we will discuss two extreme cases. In each case we assume we have made M measurements on each of N different spots. In total we therefore have M times N measurements.

Let us first consider the case that all $M \times N$ measurements are independent, i.e. random noise is the main source of the uncertainties. In this case we can neglect the covariance term in the uncertainty propagation. Covariance is discussed in more detail in Chapter 4. The resulting combined SD s_C of the population expressed as a function of the external and internal uncertainties is given by:

$$s_C^2 = \frac{M(N-1)s_I^2}{MN-1} + \frac{N(M-1)s_E^2}{MN-1} \tag{3.6}$$

and the SD of the mean is given by:

$$s_\mu^2 = \frac{s_C^2}{MN} = \frac{M(N-1)}{MN-1} \frac{s_I^2}{N} + \frac{N(M-1)}{MN-1} \frac{s_E^2}{M} \tag{3.7}$$

In most cases this value will be much smaller than the external precision alone, because more information is available on the stability of the signal.

A different situation arises when the external SD is significantly larger than the internal SD, *i.e.* when the difference between two burns is significantly larger than the variations during a single burn. In this case the m measurements performed during the averaging process cannot be considered as independent (if the first value is high, all others in the same burn will be high), and acquiring many data points on the same spot does not improve the reliability of the data. Only when many different spots are measured will the uncertainty of the mean decrease.

The combined uncertainty is then:

$$s_C^2 \approx s_E^2 \tag{3.8}$$

and the uncertainty of the mean of the $M \times N$ measurements is given by:

$$s_\mu^2 = \frac{s_I^2}{M} + \frac{s_E^2}{N} \approx \frac{s_E^2}{N} \tag{3.9}$$

The calculation of internal and external precisions is illustrated for several elements in brass CTIF L6 in Table 3.3. Nine measurements were made for each element: three repeat measurements in each of three craters. The internal precision was determined by averaging the SD for each set of three repetitions. Note that variations in internal precision can help identify spikes. The external precision was determined by first taking the means for each crater and then taking the SD of these mean values.

Separation of the data into internal and external precisions helps identifying where the largest uncertainty lies (see Table 3.4). Here the internal precision was much lower than the external precision and no spikes were detected, so the combined SD was determined using Equation 3.9. The RSDs of the mean were <2% as expected.

Table 3.3 *Measurements I1–I3, I4–I6 and I7–I9 are three repetitions on three different burns (craters) on brass CTIF L6*

Line	I1	I2	I3	I4	I5	I6	I7	I8	I9
Cu 225	1.974	1.972	1.970	1.958	1.952	1.946	1.945	1.940	1.935
Ni 341	1.285	1.289	1.290	1.261	1.263	1.263	1.260	1.262	1.263
Si 288	3.624	3.633	3.643	3.534	3.535	3.533	3.480	3.483	3.477
Al 396	0.436	0.437	0.437	0.427	0.427	0.4276	0.418	0.418	0.417

Table 3.4 *Results for brass CTIF L6, for measurements I1–I9 in Table 3.3*

Line	Mean	s_I	s_E	s_C	s_μ	RSD_{Mean} (%)
Cu 225	1.9547	0.0043	0.0162	0.0145	0.0093	0.48
Ni 341	1.2708	0.0018	0.0150	0.0131	0.0087	0.68
Si 288	3.549	0.0045	0.0778	0.0675	0.045	1.3
Al 396	0.4272	0.0003	0.0096	0.0083	0.0056	1.3

5 Memory Effects

Memory effects are not a serious problem for GDOES, as they have not been commonly observed. Due to the design of the discharge source, sputtering should only occur on the sample surface and not on the anode. The potential difference between the plasma and the anode surface is normally too small to create sputtering. If the anode-to-sample gap is too large (>0.4 mm), some sputtering of the copper cathode block can occur in DC operation on old sources without a ceramic ring. The cathode block is normally an insulating material (*e.g.* ceramic) in RF operation.

Part of the sputtered material is deposited inside and on the front face of the anode, part of the material is redeposited on the edges of the crater and some pumped away with the argon (see Figure 3.10). The discharge tube is reamed regularly to remove material that has been deposited on the anode surface.

A potential risk for memory effects exists when organic or other volatile material is analysed. When such material is deposited inside the discharge chamber, beyond the reamed area of the anode, it may evaporate during the next analysis and a memory effect could be observed. This is more likely when material such as paint is analysed at high power.

6 Voltage, Power, Current, Pressure

For some years now, there has been a lively discussion about voltage, current, power and pressure, *i.e.* about the discharge parameters, and their influence on results. In particular, the discussion has focused on the influence of the discharge parameters on emission yield, the probability that an atom or ion in the discharge will emit a photon of a given wavelength. We will discuss this further in Chapter 11. Here the

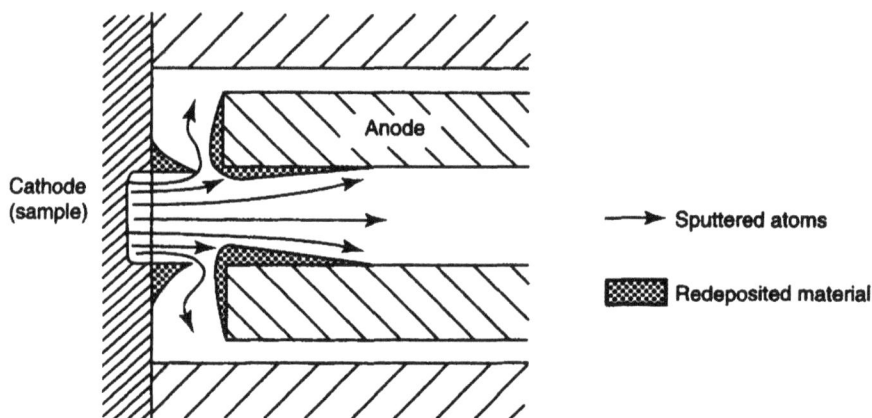

Figure 3.10   *Redeposition of sputtered material, onto the sample surface, onto the crater edges, and the anode*[4]
(Reproduced with permission from M. Bouchacourt and F. Schwoehrer, in *Glow Discharge Optical Emission Spectrometry*, R. Payling, D.G. Jones and A. Bengtson (eds), John Wiley & Sons, Chichester, 1997, 51–3)

aim is simply to give some advice on how to optimise these parameters for reliable analytical results.

For a constant source design, not all of these parameters can be fixed independently. When voltage and current are fixed the discharge power, which is the product of voltage and current, is also fixed. But then the pressure would not be fixed if we varied the sample matrix, as we would need to adjust the pressure to maintain the required current for a given voltage. Similar arguments are true for any other combinations of parameters, such as constant power and pressure, where the ratio of current to voltage will change when the sample matrix is changed.

Figure 3.11 shows the variation of DC voltage and current for different pressures for a nickel sample. For constant pressure, as the voltage is increased the current also increases, though not linearly; increasing pressure gives a lower voltage for a given current.

Plasmas, in general, are fairly difficult to describe, and even more difficult to understand. This is particularly true for the plasma used in GDOES. Do not be too concerned if you do not understand the plasma in minute detail, few, if any, of us really do. In GDOES, we are working at pressures of some hundreds of pascal, where neither the approximations for low pressure, *i.e.* few collisions, nor the approximations for high pressure, *i.e.* one big collision (studied *e.g.* in fusion processes) are valid. In addition, the plasma is confined to a small but not tiny volume of a few tens of cubic millimetres, where plasma models are difficult to implement.

It should also be kept in mind that exact measurements of the various plasma parameters are not easy. This is particularly true for pressure and some RF parameters. An important plasma parameter, *i.e.* gas temperature, is usually not even measured at all. So let us begin by contradicting ourselves and imagine we can study the influence of each parameter independently.

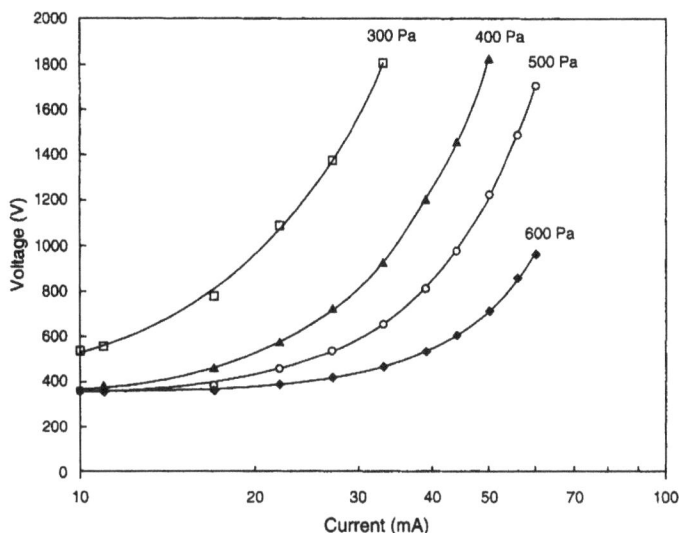

Figure 3.11 *Relationship between current, voltage and pressure in DC operation on a nickel sample[5]. In RF operation, the actual values would be different but the general trends would be similar*
(Reproduced with permission from M. Bouchacourt and F. Schwoehrer, in *Glow Discharge Optical Emission Spectrometory*, R. Payling, D.G. Jones and A. Bengtson (eds), John Wiley & Sons, Chichester, 1997, 62–6)

Voltage

The voltage drop appears essentially only in the cathode dark space (CDS), which means the small distance between the cathode, *i.e.* the sample, and the plasma glow. The glow region, called the negative glow (NG), being basically free of electrical field, does not participate in the applied voltage. As a consequence, and neglecting all other effects, increasing the voltage should lead to a higher energy of the ions bombarding the sample surface, hence to an increased sputtering rate, and a higher average energy of electrons being accelerated towards the plasma glow, allowing atoms and ions to be excited to higher electronic states.

Increase in the voltage shortens the CDS and lengthens the NG.[6] It also decreases the emission yield of many spectral lines of analytical interest, because of the reduced collision cross-section with the higher energy electrons. But the increase in sputtering rate is greater than the decrease in emission yield, so the intensity still increases with increased voltage.

Current

At high voltage but very low currents, the plasma is first a Townsend dark discharge and then a normal discharge, and as the current is increased it changes to an abnormal discharge and finally to an arc discharge. Figure 3.12a shows how these transitions

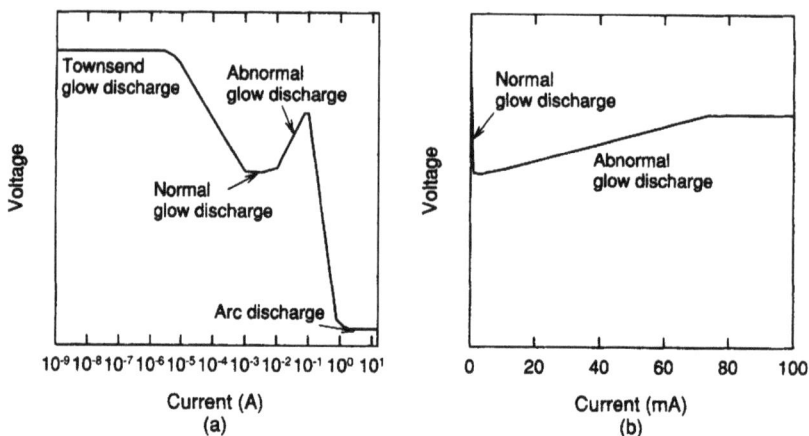

Figure 3.12 *General relationship between current and voltage over a wide range of conditions: (a) as is commonly represented, following Penning;[7,8] and (b) on a linear scale to emphasise that the abnormal region dominates at currents below 100 mA*

are normally presented, with a logarithmic current scale. But when a linear scale is used, as in Figure 3.12b, the importance of the abnormal discharge from 0 to 100 mA is more readily seen.

The plasma used in GDOES is the so-called abnormal discharge. It behaves like a resistor in that current and voltage increase together, though they are not strictly proportional as in a perfect resistance. It is unlike a normal discharge where the discharge voltage and current are nearly independent of each other.

In abnormal discharges the entire cathode surface is covered by the discharge. An increase in the discharge current leads to an increase in the current density. It will probably also lead to a higher density of charged particles in the discharge. Increasing the current has little effect on the size of the CDS and NG,[6] but increasing the number of excited particles in the plasma increases both the sputtering rate and the emission yield. Therefore, current has a large effect on intensities.

When the discharge current is reduced, the current density is also reduced. When the discharge current is sufficiently reduced, we approach the normal discharge regime, the current density is no longer uniform over the cathode surface. We find a higher current density in the centre of the cathode leading to non-uniform sputtering and the formation of a U-shaped crater, usually referred to as a convex crater. At 'soft' conditions, *i.e.* in the normal discharge region, therefore, the sample surface is not fully covered by the plasma and sputtering is concentrated near the centre of the analysis area. The current increases rapidly with increasing voltage. At the other extreme, *i.e.* at 'hard conditions', approaching an arc discharge, the plasma becomes very unstable. Before this the crater becomes concave in shape as the field intensifies at the edges of the crater. Various examples of these crater shapes are shown later.

Table 3.5 *Effect of increasing pressure for constant power, current or voltage*

Power	Current	Voltage	Impedance	Sputtering rate	Intensity	Crater shape
Constant	↑	↓	↓	↑	↑	Convex to concave
↓	Constant	↓	↓	↓	↓	Convex to concave
↑	↑	Constant	↓	↑	↑	Convex to concave

Power

The main effect of increasing the power seems to be an increase in sputtering rate. The increased sputtering leads to an increase in the luminescence of the discharge. The emission yields for many spectral lines seem to be nearly independent of the plasma power. While it is desirable to work at high power, as more material is removed per unit time, the power dissipated in the plasma leads to an increase in temperature, particularly of the sample surface. Depending on the heat resistance of the sample, there is a natural limit to increasing the discharge power caused by the plasma becoming unstable through evaporation of the sample surface.

Pressure

The argon gas pressure and temperature determine the argon atom density in the plasma chamber. With increased gas density, the probability of collisions is increased. This will lead to an increase in ion density and to a decrease in average energy, as the mean free path for charged particles is reduced. Pressure seems to have little direct effect on sputtering rates. Its effect on emission yields is less easily established, as the effect depends on the materials and the spectral lines studied. For many lines, however, increasing pressure seems to decrease the emission yield slightly.[9]

It is worth remembering that an increased number of collisions will lead to a decrease in the plasma impedance, *i.e.* when the pressure is increased at constant power, the voltage will drop and the current will increase. Increase in the pressure will also shorten both the cathode dark space and the negative glow.[6] When the discharge is operated at constant power, current or voltage, increasing the pressure will change the source impedance, sputtering rates and crater shape, as indicated in Table 3.5.

7 Optimising Plasma Parameters for Bulk Analysis

Before speaking about optimising parameters, we should first establish what the criteria are for assessing the optimum. For bulk analysis we want to achieve the following:

- short analysis time
- low detection limits for trace elements
- high reproducibility for major alloy elements

Table 3.6 *Recommended conditions for bulk analysis (actual values will vary from instrument to instrument and with anode diameter)*

Material	Power (RF)	Current (DC)	Pressure
Al, Al–Si	High	High	High
Brass	Medium	Medium	Medium
Pb, Sn	Low	Low	Medium
Steels	High	High	High
Zn	Medium	Medium	Medium
Ceramics	High	–	Low to medium
Glass	Low to medium	–	Low to medium
Polymers	Low	–	Low

As a consequence we will probably tend to choose high power and high pressure, or equivalently, high voltage and current. But, as noted earlier, there are limits. Materials with a low melting point, such as tin, lead alloys, zinc, *etc.*, cannot be run at high power because their surfaces will melt and the discharge will become unstable. Typical conditions for some common types of materials are given in Table 3.6.

In terms of analysis time, high power is beneficial as more material is removed from the sample surface, therefore the sample surface is cleaned much faster, and pre-integration times can be reduced. For trace elements it is also clear that high power is beneficial as more material is removed from the sample to the plasma, and so signals are higher during integration.

For stability, we need to find the optimum conditions by repeating the analysis a few times at different conditions. Running the analysis in time resolved (depth profiling) mode shows very quickly how stable the discharge is over time and how much time is needed to find stable conditions. Usually, the best conditions for bulk analysis are when the crater bottom is slightly U-shaped, *i.e.* with high pressure or high current.

8 Optimising Plasma Parameters for Depth Profiling

When optimising parameters for surface analysis or depth profiling, we want to achieve the following:

- high reproducibility for major and minor elements
- good depth resolution
- low signal noise, but with sufficient number of points to see the features of interest

As a consequence we will probably choose a power that does not overheat the sample (or crack the sample in the case of glass), a low to moderate pressure to improve depth resolution and a short integration time especially at the beginning of the acquisition. Typical conditions for some common types of materials are given in Table 3.7.

Table 3.7 *Good conditions for typical applications in depth profiling (RF or DC, actual values will vary from instrument to instrument and with anode diameter)*

Coating	Power	Pressure
Al, Al–Si	Medium	Medium
Sn	Low	Low to medium
Steel surface	Medium	Medium
Zn–Al	Low to medium	Medium
Ceramics	Medium	Low
Glass	Medium	Low
Polymers	Medium	Low

Crater Shapes

Crater shapes are an essential factor when dealing with depth profiling by GDOES. To get good depth resolution the crater bottom must be flat, or just slightly concave at the edges.[10]

One of the nice features of the analytical GD source is its ability to produce flat craters. But depending on the discharge conditions and the samples, the craters are not always flat. Intensive studies have been performed in various laboratories on the conditions for achieving flat crater bottoms in different materials. Typical conditions for DC operation are shown in Figure 3.13. RF operation shows similar trends, though because RF current is difficult to measure, the *x*-axis would be power.

As a general rule we find that moderately high pressure (low voltage and high current or low impedance) leads to a U-shaped (convex) crater, *i.e.* there is more sputtering in the centre of the burn spot than at the edges. At the other extreme, when the pressure is too low, the crater will be concave, *i.e.* we find faster sputtering at the edges of the crater than in the centre.

A 'profilometer' is an efficient means for optimising the crater shape. Profilometers are described in more detail in Chapter 6. A bulk sample or coated sample is sputtered, the crater shape is measured with the profilometer and the discharge conditions are adjusted until a flat crater is obtained. In RF operation, this usually involves selecting a power and changing the pressure to optimise the crater shape. In DC operation, a current or voltage is usually selected and again the pressure is adjusted to optimise the crater shape.

If a profilometer is not available, it is sometimes possible to judge the crater shape by eye, especially when a coated material is used and the discharge is stopped at the interface. Alternatively, the pressure can be adjusted to minimise the measured width in a depth profile of the interface between a coating and a substrate. An example is shown in Figure 3.14, where depth profiles were recorded from a commercial Zn–Ni coated steel at four different argon pressures. The optimum pressure here appears to be about 580 Pa.

In all cases, a material similar to the coating to be analysed should be used, as the optimal conditions for flat craters depend on the material being analysed. For

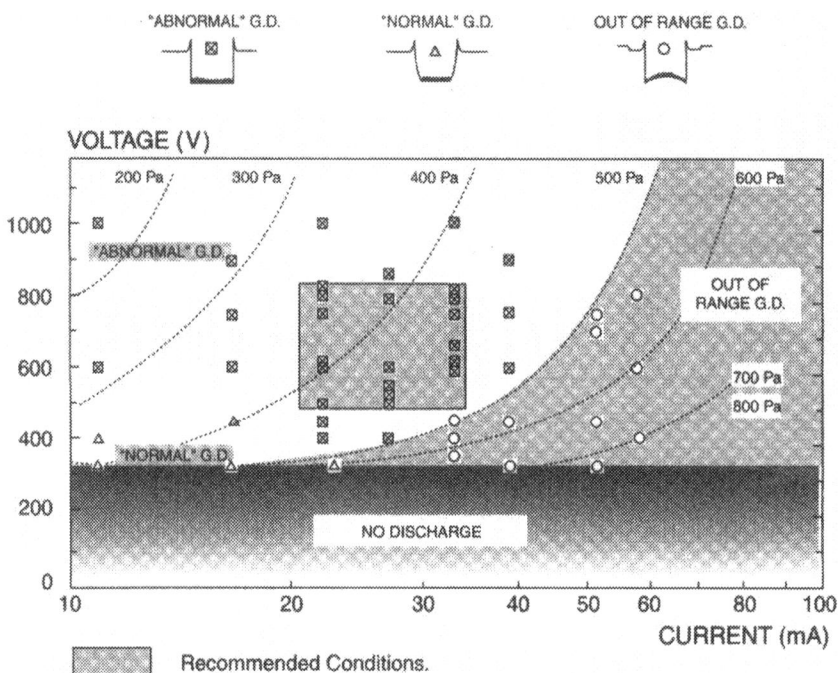

Figure 3.13 *Dependence of crater shape on source conditions in DC operation*[11] *(4 mm an-ode, carbon steel)*
(Reproduced with permission from M. Bouchacourt and F. Schwoehrer, in *Glow Discharge Optical Emission Spectrometry*, R. Payling, D.G. Jones and A. Bengt-son (eds), John Wiley & Sons, Chichester, 1997, 62)

coated materials often the crater shape is checked at the interface. When no such material is available for optimisation, material showing similar 'discharge' behaviour should be used. By similar discharge behaviour, we mean that, for a given pressure, the discharge impedance should be similar for the test sample and the sample to be analysed. For example Cr, Ni and Fe have similar behaviour, while Zn and Al are similar to each other but different from Cr, Ni and Fe. This topic is discussed further in Chapter 11.

Depth Resolution

Depth resolution in sputtering techniques such as GDOES and SIMS (secondary ion mass spectrometry) has been defined by ISO as the distance between the 84% and 16% points in a depth profile across a sharp interface divided by the depth at 50% (in sputtering time or depth).[12] The numbers 84% and 16% were chosen to represent the width of a normal distribution. The measured profile is assumed to be the convolution of a depth function, resembling a Gaussian profile, and a sharp interface, idealised by a step function. The convolution of these two functions is an error function, as shown

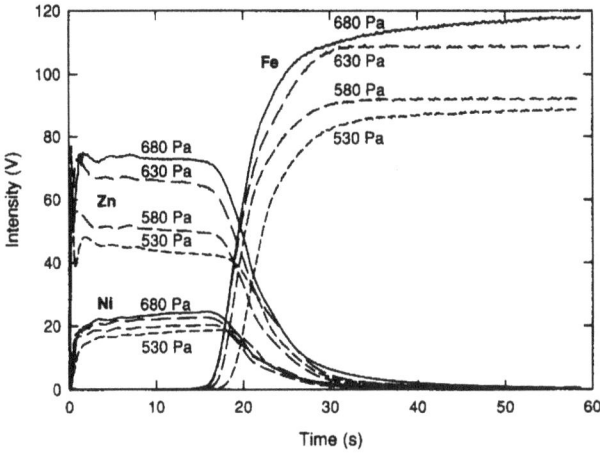

Figure 3.14 *Variation in depth resolution for a Zn–Ni coated steel at different argon pressures*

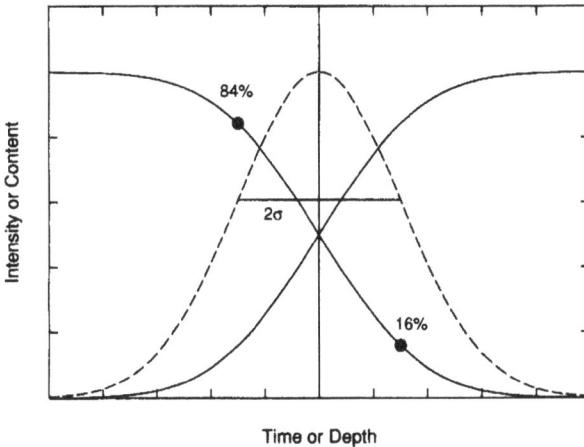

Figure 3.15 *The 84% and 16% points used in the definition of depth resolution and how they relate to the width (2σ, i.e. ±1σ) of the normal distribution (dashed). The solid lines are error functions.*

in Figure 3.15. Even if the actual profile is not a pure error function, the 84% and 16% are a sensible way to standardise the measurement of depth resolution.

The Gaussian profile assumes that sputtering is a statistical process that produces a flat crater bottom with random variations. In real GD depth profiles, variations in crater shape will distort the measured profile. Variations in sputtering rates in different layers will also distort the depth profile. For example, a slow sputtering coating on a fast sputtering substrate (*e.g.* aluminium on brass or polymer on metal) will elongate the measured profiles for elements in the coating, on both sides of the true profile, while a fast sputtering coating on a slow sputtering substrate (*e.g.* zinc on steel) will shorten the measured profiles for coating elements. See Chapter 9 for a more detailed discussion of this effect.

In some cases the depth profile departs so much from an 'ideal' error function that one or both 84% and 16% points are not well defined. An alternative has been suggested based on the inverse maximal slope near the interface. For a compositional depth profile, the depth resolution is then given by[13,14]:

$$\Delta z = \Delta c / \left(\frac{dc}{dz} \right)_{max} \qquad (3.10)$$

For a qualitative depth profile, depth z and composition c could be replaced by time and intensity, respectively.

The depth resolution that can be obtained by GDOES depends very much on the nature of the sample. Shimizu *et al.* have found that for oxide films on highly polished aluminium the depth resolution is a few nanometres, comparable to SIMS.[15] On SiO_2 coated silicon wavers the depth resolution is about 2% of the total thickness down to several hundred nanometres.[16] These are extreme cases, however, and usually the depth resolution is not as good as this, principally because of the nature of the samples being studied rather than the limitations of the GD source. For example, poorer depth resolution may be found because the interface between two layers is not well defined due to roughness, migration processes or crystalline structure. Variations in coating thickness over the analysed area will also deteriorate the measured depth resolution. As a rule of thumb for industrial materials, depth resolution is about 15% of the depth.

Generally, the mean roughness of the bottom of the crater increases with sputtered depth and can be as much as 50% of the depth for highly structured (granular) materials such as brass.[17] The overall crater shape, however, seems to change little with depth, see Figure 3.16; the convex or concave sides growing in proportion to the depth of the centre of the crater.

9 Sequence for Optimising Parameters

The following steps are recommended for setting up source and analysis parameters:

1. select suitable flush time (if available)
2. select suitable pre-burn (pre-integration) time
3. select integration time for calibration and bulk analysis
4. select suitable source mode, *e.g.* constant pressure and applied power, or constant current and voltage
5. select suitable power, either applied power or current × voltage
6. adjust other source parameter(s) for optimum signal (bulk) or optimum crater shape
7. adjust detectors for expected signals

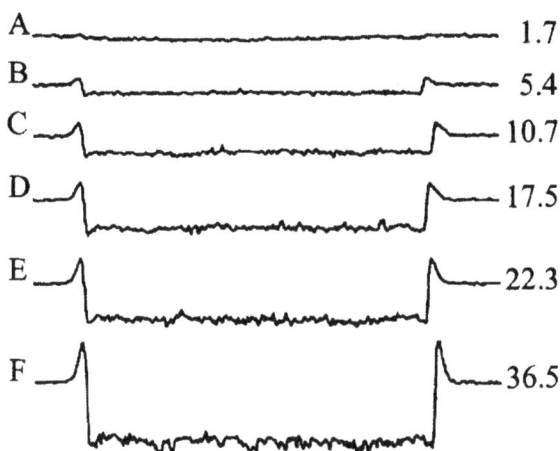

Figure 3.16 *Crater profiles after various depths[17] (μm)*
(Reproduced with permission from A. Quentmeier, in *Glow Discharge Optical Emission* Spectrometry, R. Payling, D.G. Jones and A. Bengtson (eds), John Wiley & Sons, Chichester, 1997, 300)

References

1. D.W. Allan, N. Ashby and C.C. Hodge, HP Application Note 1289, *The Science of Timekeeping*, Hewlett Packard, 2002.
2. C. Audoin and B. Guinot, *Les fondements de la mesure du temps*, Masson, Paris, 1998, 70–7.
3. M. Winchester and J.K. Miller, *J. Anal. Atom. Spectrom.*, 2001, **16**, 122.
4. M. Bouchacourt and F. Schwoehrer, in *Glow Discharge Optical Emission Spectrometry*, R. Payling, D.G. Jones and A. Bengtson (eds), John Wiley & Sons, Chichester, 1997, 51–3.
5. M. Bouchacourt and F. Schwoehrer, in *Glow Discharge Optical Emission Spectrometry*, R. Payling, D.G. Jones and A. Bengtson (eds), John Wiley & Sons, Chichester, 1997, 62–6.
6. A. Bogaerts and R. Gijbels, *Spectrochim. Acta B*, 1998, **53**, 1.
7. F.M. Penning, *Philips Technical Library*, Einhoven, 1957.
8. K. Wagatsuma, in *Glow Discharge Optical Emission Spectrometry*, R. Payling, D.G. Jones and A. Bengtson (eds), John Wiley & Sons, Chichester, 1997, 167–75.
9. R. Payling, *Surf. Interface Anal.*, 1995, **23**, 12.
10. K. Shimizu, H. Habazaki, P. Skeldon, G.E. Thompson and R.K. Marcus, *Surf. Interface Anal.*, 2002, **31**, 869.
11. M. Bouchacourt and F. Schwoehrer, in *Glow Discharge Optical Emission Spectrometry*, R. Payling, D.G. Jones and A. Bengtson (eds), John Wiley & Sons, Chichester, 1997, 62.
12. ISO 14707, *"Glow discharge optical emission spectrometry (GD-OES)—Introduction for use"*, 2001.
13. S. Hofmann, *Surf. Interface Anal.*, 1999, **27**, 825.
14. V. Hoffmann, *Internal. Symp. on Surface Analysis by GDOES Keio University, Japan*, November 2003.

15. K. Shimizu, G.M. Brown, H. Habazaki, K. Kobayashi, P. Skeldon, G.E. Thompson and G.C. Wood, *Surf. Interface Anal.*, 1999, **27**, 24.
16. T. Nelis and R. Payling, in *Surface Analytical Measurements in Materials Science*, 2nd edn., D.J. O'Connor, B.A. Sexton and R.St.C. Smart (eds), Springer-Verlag, Berlin, 2003, 553.
17. A. Quentmeier, in *Glow Discharge Optical Emission Spectrometry*, R. Payling, D.G. Jones and A. Bengtson (eds), John Wiley & Sons, Chichester, 1997, 300.

CHAPTER 4

Calibration

1 Introduction

GDOES is a comparative analytical method and therefore needs calibration. In fact, calibration is the most important part of the preparation for analysis as the final analytical result can never be more reliable than the calibration.

The concept of calibration is extremely simple. By measuring the emitted intensities for a set of samples with known composition, a 'relationship' between measured intensities and composition is established. The relationship is then used to assign the chemical composition of an unknown sample on the basis of measured intensities.

As often with very simple concepts, some care must be taken when implementing in order to avoid mistakes. The main problem of calibration is in the choice of the 'relationship' between composition and intensity.

ISO Standards

As described in the ISO 9000 standard, quality management systems require specific care for measurements, measurement equipment and related uncertainties. ISO standards 10012-1 and 17025 treat the requirements for quality assurance when measurements are performed for a third party.[1,2] In particular, these standards are applicable to measurement equipment used for the confirmation of specifications in a supplier–customer relationship. They are also directly applicable to analysis laboratories supplying calibration services. Some of the requirements, such as organising audits, designating a person responsible for the measurement equipment, or assuring operators are properly trained for the job, concern mainly the administration of the laboratory. They are not the subject of this book. Other requirements, however, are directly linked to our subject, especially to calibration.

To comply with the ISO quality management system, the laboratory must not only be sure that the measurement equipment is able to perform the required measurements with the requested precision, but also be able to demonstrate this in reports. For an analytical tool such as a GDOES instrument, this means, for example, if the analytical problem is to determine whether the Zn content in a brass sample is between 19.1% and 19.4%, the precision of the instrument and the accuracy of the analysis procedure must be sufficient to satisfy this requirement. If we are thinking, instead, about depth profiling, it must be possible, for example, to demonstrate that the estimate of the total layer thickness is better than $\pm x$ μm.

Compliance requires:

- a proper treatment of measurement uncertainties
- protection of calibration features of the equipment
- creation and maintenance of links between the final measurement and national and international standards, usually through certified reference materials (note that there is also provision in ISO standards for the use of consensus materials, *i.e.* materials agreed to by all parties concerned[2])

An estimation of uncertainties is not only interesting because of ISO standards, it also helps guide the operator in making the right choices when using the instrument. But a thorough estimation of uncertainties has not yet become a normal practice throughout the analytical world. We have therefore dedicated a significant part of this book to this often overlooked aspect of GD analysis, beginning with this chapter.

Sometimes it is still the dream of analysts to make a measurement as precise as possible. But when we look at the requirements of most modern laboratories, measurements must be made with a specified precision. To distinguish two different types of alloys, it is not necessary to measure all elements with relative precisions in the 0.1% range. An understanding of how the uncertainties of the final results can be estimated and defining the major sources of uncertainties do help save time and eventually money.

Some Definitions Used in Calibration and Analysis

Standard is a document containing 'technical specifications or other precise criteria to be used consistently as rules, guidelines, or definitions of characteristics, to ensure that materials, products, processes and services are fit for their purpose'.[3] The intention is that if the operator follows the standard, then they should achieve the stated outcome(s) in the document. Historically, in chemical analysis, reference materials have also been called 'standards', but this use is now discouraged because of the possible confusion between documents and materials. The first ISO standard for GDOES is an 'Introduction for use'.[4]

Precision is an estimate of variation in a series of measurements, usually expressed as a standard deviation. The terms internal and external precision have been defined in Chapter 3. In chemical analysis, there are three important ways of estimating precision: repeatability, intermediate precision and inter-laboratory reproducibility.[5]

Repeatability is the variation in a measurement made by a single operator on the same instrument over a short time.

Intermediate precision is the variation in a measurement where one or more factors, such as the operator, instrument, or time, are changed within a single laboratory.

Reproducibility (inter-laboratory) is the variation in a measurement made by different laboratories on the same sample.

Uncertainty is a parameter associated with a measurement that 'characterises the dispersion of the values that could reasonably be attributed to' the measurement.[6]

Accuracy is the closeness of agreement between a measurement and its accepted value. It should not be confused with precision.

Error is the difference between a measurement and its accepted value.

Traceability is 'the establishment of an unbroken chain of comparisons to stated references'.[7,8] Traceability concerns the results of measurements and the values of standards, and normally involves references to national and international standards. All comparisons must be documented, including uncertainties. In GDOES, traceability concerns the values of standards used in measurements and analysis results. In bulk analysis, the principal concern is the elemental content of samples (normally assessed using certified reference materials (CRMs)). In compositional depth profiling, it also concerns the estimated depth (normally compared with profilometry or coating mass). Establishing traceability involves a description of the instrument and the standard method followed, the result and its uncertainty, specifications of relevant national/international references, and internal assessment programmes to establish the status of the instrument, standards and references. For some GDOES laboratories, establishing traceability may be crucial to their performance. For most GDOES laboratories, traceability is a valuable way of checking that their results do not wander too far from internationally accepted values.

Variance is the second-order moment of a probability density distribution for a random variable x. (The mean is the first-order moment.) It is a measure of the dispersion of values around the mean, μ, with a probability distribution, $f(x)$:

$$Variance(x) = \int (x - \mu)^2 f(x) \, dx \qquad (4.1)$$

Covariance is defined for two random variables, x and y, having a joint probability distribution, $f(xy)$, with μ_x and μ_y being the expected (mean) values for the variables x and y respectively. It is a measure of the extent to which x and y are correlated. The covariance is given by:

$$Cov(x, y) = \iint (x - \mu_x)(y - \mu_y) f(x, y) \, dx \, dy \qquad (4.2)$$

The covariance can vary from a large positive number, for two variables increasing or decreasing together, to a large negative number, when one increases as the other decreases, and equals zero when the two random variables are independent.

Table 4.1 *Elements, wavelengths and compositions of interest in our brass calibration*

Element	Wavelength (nm)	State	Content range (mass%)
Cu	224.700	II	55–95
Zn	334.502	I	5–44
Mn	257.611	II R	0.00–4
Al	396.152	I r	0.01–3
Ni	341.476	I r	0.02–3
Pb	220.356	II	0.02–3
Si	288.158	I	0.04–2
Fe	371.994	I R	0.03–2
Sn	317.505	I	0.01–1.5
P	178.283	I R	0.02–0.2

2 Preparation for Calibration

The first step in preparing for a calibration of an analytical instrument is to define the purpose of the calibration:

- elements to be covered
- covered composition ranges
- required precision
- frequency of use

After considering these criteria, it will become easier to make the right choice about how much effort must be spent in calibrating the instrument for the given application.

To illustrate, consider we are making a new brass calibration. The elements and composition ranges of interest are shown in Table 4.1. The composition ranges are broad, as we are going to cover the full range for each element, we will need many calibration samples, and we want to do a good job. Also shown in the table are the wavelengths chosen on the instrument available and their excited states, more about these will be discussed later.

Selection of Calibration Samples

As mentioned, the calibration process is crucial to the reliability of the analytical results obtained with the instrument. Clearly, therefore, calibration materials (see Figure 4.1) should be selected with care. Unfortunately, CRMs are usually expensive and not always available in unlimited quantities. The following recommendations should therefore be followed whenever possible, but for various reasons, such as time or money, it will not always be possible to follow them strictly. It is therefore also a purpose of this chapter to provide some understanding of the calibration process, so users will be able to make their own choices and defend them.

Recent CRMs are usually supplied with certificates of the chemical composition and with estimated uncertainties in the certified values. These estimated uncertainties

Figure 4.1 *Selection of steel CRMs, polished ready for GDOES calibration*

not only allow an estimation of the quality of the standards but also are an important source of information for estimating uncertainties in any calibration using them.

As mentioned in Chapter 3, GDOES uses only a small amount of material for the actual analysis. This implies that any material used for calibration should be very homogeneous. When we speak here about homogeneity, we are not referring to large bars of material that must have the same average composition throughout, but a small sample used for calibration. Once you are sure about the stability of your instrument, you can check the homogeneity of calibration samples simply by repetitive analysis of the sample. If the relative standard deviations (RSDs) of the intensities for major and minor elements in ten consecutive impacts are worse than those obtained with other samples, the variation is probably due to the sample. You will be surprised by the differences in homogeneity of different samples.

Number of Calibration Samples

The number of samples needed for calibration is still very much discussed. Given that the mathematics used for such calculations is now well developed and accepted, except for minor details, the simplest explanation for this ongoing debate is different priorities. Uncertainties in calibration tend to decrease as the number of samples is increased (see Figure 4.2) but the time and cost of calibration both increase. It is therefore extremely difficult to give a fixed value for the number of calibration samples that is necessary and sufficient for a good calibration. In fact the number depends on the expected quality and the covered composition ranges.

One of the interesting features of GDOES is the fact that spectral interferences are fairly rare and that matrix effects are reduced, in comparison with other techniques such as Spark source OES. This means a calibration can often be done with fewer calibration samples or over larger composition ranges, but usually not both.

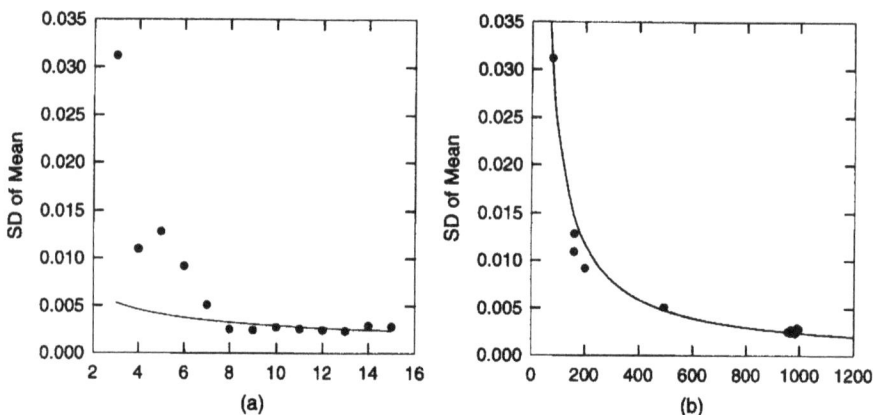

Figure 4.2 *Standard deviation of the mean (centroid) of calibration curve for Al 396 in brass, as a function of (a) number of calibration samples and (b) square root of sum of weights, where the samples were selected at random and added one by one to the calibration curve. Solid line is the expected trend. Weighted regression is discussed later*

As rule of thumb, we suggest a minimum of six calibration samples per calibration curve in a standard case, for a linear calibration curve with only two regression parameters (slope and intercept).

When calibrating alloys with high contents of different elements and when significant interferences are to be expected (*i.e.* significant content of elements showing a rich line spectrum, *e.g.* Co, Mo, W, *etc.*), the number of samples should be increased. This is the only way to be sure that interferences are correctly taken into account.

When large composition ranges are to be covered, again the number of calibration samples should be increased. Though matrix effects are reduced in GDOES, they cannot simply be ignored. Increasing the number of calibration samples is again the only way to detect and possibly correct these effects.

When non-linear calibration curves are used, again the number of calibration samples should be increased, because calculating three parameters (*e.g.* in a second-order polynomial) with only six data points does not make much statistical sense. Choosing at least three data points for each repression parameter is certainly not excessive, but rather at the low end of what is sensible.

Except in special cases, binary samples should be avoided in calibration, as they contain too little elemental information, but require the same measurement time. But certainly, they are better than nothing.

Having seen all the exceptions, when the number of calibration samples per spectral line should be more than six, you will not be surprised to see calibrations using a total of more than 20 different samples.

It should also be mentioned that calibration samples of different suppliers should be used. Although this does not improve the appearance of a calibration curve, because sets of CRMs prepared by different suppliers do not always agree well with each other, the resulting calibration function will give a more balanced picture of the real situation. (This is, of course, not true if you want to reproduce the same figures as the manufacturer of one set of CRMs.)

Composition Ranges

The samples included in the calibration process should cover the entire composition range that will be used for analysis. Although extrapolation of calibration curves is possible in GDOES, because of its mainly linear calibration curves, and sometimes necessary, due to the lack of suitable reference materials, extrapolation should be avoided whenever possible because of the increased uncertainties that result.

The debate, within the analytical community, on whether it is better to use a wide range of calibration samples with compositions spread evenly over the composition range, or whether it is better to use calibration samples in clouds near the minimum and maximum compositions is still going on. A general answer is not possible, nor sensible.

Using several samples with minimum or maximum composition will allow a reduction of the uncertainties in determining the slope and intercept of a linear calibration curve. This configuration also leads to a reduced correlation between the two regression parameters.

On the other hand, having calibration samples spread evenly over the entire composition range will verify the linearity (or non-linearity) of the calibration curve. It will allow easier detection and correction of some matrix effects. It also means that calibration includes samples with similar compositions to those being analysed, a well-established laboratory practice.

When the spectral line used for analysis is known to be linear for the given application and free from interferences and matrix effects, it would be beneficial to include calibration samples at the two extremes of the composition range. If not, it is advisable to spread them over the entire range.

The wider the range used, the more demanding the calibration and the more difficult it is to maintain excellent figures of merit, but the more useful are the results analytically. It is possible, in GDOES, to obtain linear calibrations over two, three, or more orders of magnitude. When the analytical range is wide, it is difficult to see what is happening at low compositions on linear–linear plots. The use of log–log plots (see Figure 4.3) for calibrations with wide composition ranges is recommended.

Covariance between the different regression parameters may significantly reduce the reliability of the final regression curve. Covariance in the fit parameters appears when the columns or lines of the fit matrix are linearly independent. To avoid or at least to reduce this covariance, the calibration samples must be chosen carefully.

In 'analytical' terms, to decouple the constant term (intercept) and the linear term (slope), the compositions of the elements in the calibration samples should be widely spread. Imagine the reverse situation: we are using 10 samples, all with 10% of element A. Once we have measured their intensities, we would be unable to say whether a constant term (intercept) or a linear term (slope) would fit the data better. The two would be highly correlated.

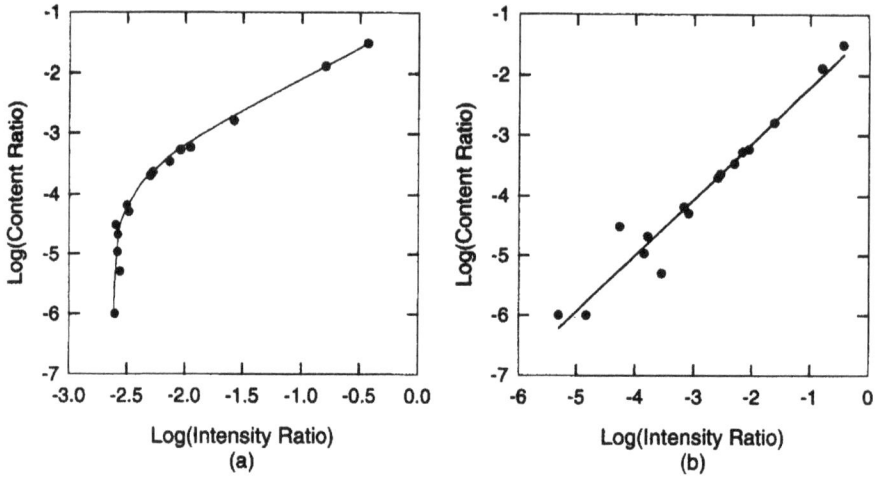

Figure 4.3 *Log–log calibration curves for Ni in Al–Si: (a) curvature as the intensity ratio approaches the background and (b) with background subtracted shows the increasing scatter evenly displaced about the regression line approaching the detection limit*

If the calibration is no longer linear and we need to include a second-order term (squared in intensity), then to reduce the correlation with the first two terms we need to add some calibration samples showing a medium composition of our element. Otherwise, we cannot distinguish between linear and squared terms. For a complete decoupling of the linear and squared terms, it would be best to add some calibration samples with negative compositions, but they have not been invented yet.

To determine a spectral interference, it is best to include samples with high and low contents of the interfering element, and to ensure that not all samples are high or low at the same time as they are high or low in the interfered element. In such a case, we would be unable to distinguish the interference from the slope of the calibration curve.

3   Calibration Function

The correct choice of calibration function is essential for successful calibration and analysis. The calibration function models the relationship between the measured intensities and the composition of the analysed sample. Given the complexity of the plasma processes in a glow discharge, it is easy to understand that the calibration function can only approximate this 'reality'. In addition, we use only a limited set of calibration samples to establish the relationship. The task is to find the best approximation, to understand its limitations and uncertainties, and finally to use it where appropriate.

Despite the complexity of the plasma, GDOES is a fairly friendly tool. In many cases the relationship between the intensities and the compositions is a straight line. Things start to become more complex when large composition ranges are to

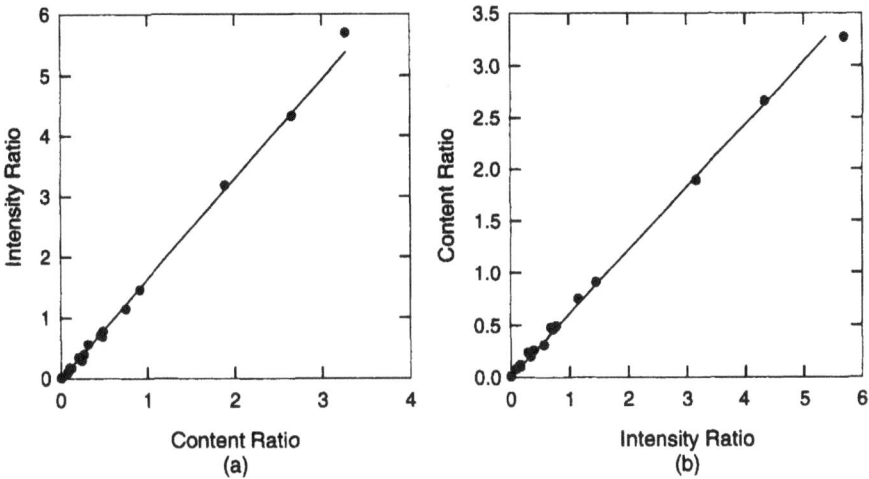

Figure 4.4 *Two ways of presenting calibration data, for Al 396 in brass, using the ratio method: (a) calibration curve and (b) analytical curve*

be covered with a single calibration function and, at the same time, high accuracy is expected. When the real physical processes are too complex to model in detail, a power series (better known as a polynomial approach) is very successful in GDOES.

Calibration Function Versus Analytical Function

To avoid confusion, we should mention two terms that are not always clearly distinguished in chemical analysis: the calibration function, where intensity is plotted versus composition, and the analytical function, where composition is plotted versus intensity. IUPAC refers to the former as an analytical calibration curve and the latter as an analytical evaluation curve.[9] These two functions are shown in Figure 4.4. It is important to realise that it is not simply a matter of how the data are plotted, this would not really change anything of importance, rather it is the selection of which variable becomes the dependent variable in the regression that matters. By convention, the dependent variable is drawn on the *y*-axis.

During the calibration process, we choose the composition and then measure the intensities. Hence we obtain a calibration function as:

$$I = f(c) \tag{4.3}$$

During analysis, we measure the intensities to determine the composition. For analysis, therefore, the calibration needs to be expressed as:

$$c = f(I) \tag{4.4}$$

Some form of inversion of the calibration function is therefore necessary or an analysis cannot be carried out. But if we adopt a pure mathematical approach, we find that inversion of regression equations cannot be performed correctly except in trivial cases because of correlations between the regression parameters.

The need to use the calibration for later analysis leads to a significant statistical problem: should intensity or composition be used as the dependent variable. Either approach will lead to a different estimate, since the axes can only be swapped in a valid statistical manner if the variables are independent.

Two different approaches are used in the analytical community:

- A calibration function is established during the calibration process. The uncertainties are assumed to be symmetrical (*e.g.* ±0.3 is symmetrical, −0.2 + 0.3 is not symmetrical). The analytical function is calculated by inverting the calibration function. The inverted uncertainties are converted to symmetrical values. During the analysis step, the chemical composition of the unknown sample is calculated using the analytical function and the measured intensities.
- The analytical function is established directly during the calibration process. This means the certified compositions of the calibration samples are recalculated from the measured intensities. The uncertainties in intensity are inverted into symmetrical uncertainties in composition. The established analytical function is used directly during the analysis step.

The calibration function is usually the one shown in textbooks on regression. It is used to establish the relationship between measurements and a proposed model. These books generally show little interest in analysis, *i.e.* in using the regression function to determine the value in an unknown sample. So while the first approach is perhaps closer to an ideal calibration, we will use the second approach, as it is the one better suited for analysis. Pragmatically, as we will show below, the differences should not be significant.

Polynomial Inversion

To see how calibration functions can be inverted, consider a calibration function that can be well approximated by a polynomial:

$$I = a_0 + a_1 c + a_2 c^2 + a_3 c^3 + \cdots \tag{4.5}$$

Shifting the intercept to the left-hand side, we get:

$$(I - a_0) = a_1 c + a_2 c^2 + a_3 c^3 + \cdots \tag{4.6}$$

We now consider that the calibration function can be inverted, and the inverse function again expressed as a power series, this time in terms of the background-corrected intensities $(I - a_0)$:

$$c = b_1(I - a_0) + b_2(I - a_0)^2 + b_3(I - a_0)^3 + \cdots \tag{4.7}$$

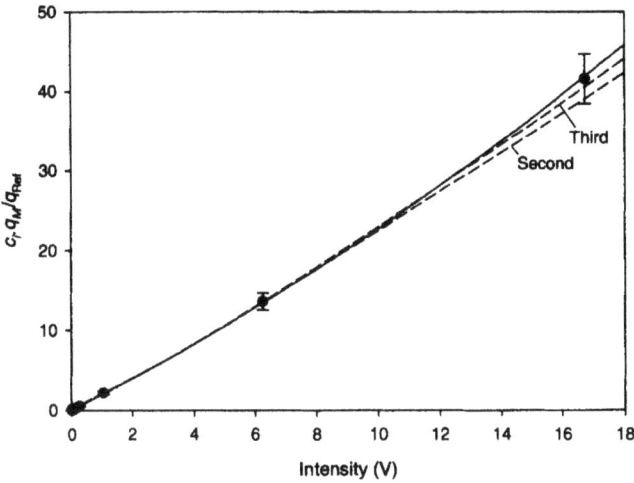

Figure 4.5 *Second-order polynomial fit using $c_i q_M/q_{ref}$ as a dependent variable (solid line), then inverted to second-order and third-order polynomials (dotted lines). Data from Ni 341 calibration described in Chapter 6*

The two sets of coefficients are related by:[10]

$$b_1 = 1/a_1 \tag{4.8}$$
$$b_2 = -a_2/a_1^3$$
$$b_3 = \left(2a_2^2 - a_1 a_3\right)/a_1^5$$
$$b_4 = \left(5a_1 a_2 a_3 - a_1^2 a_4 - 5a_2^3\right)/a_1^7$$
$$b_5 = \ldots$$

The extent to which we need to include higher powers of $(I - a_0)$ to obtain the same goodness of fit depends on the range and shape of the calibration curve. A straight line will invert exactly to another straight line, and, as Figure 4.5 shows, a gentle second-order polynomial is reasonably well represented by an inverted polynomial to second or third order.

Another problem of inversion is that the original regression parameters in Equation 4.5 are only approximate, so the values in Equation 4.7 are also approximate. We also note that at low intensities:

$$c \approx b_1(I - a_0) = b_1 I - b_0 \tag{4.9}$$

where $b_0 = a_0 b_1 = a_0/a_1$, and at high intensities, $I \gg a_0$:

$$c \approx b_1 I + b_2 I^2 + b_3 I^3 + \cdots - b_0 \tag{4.10}$$

In conclusion we can state that as long as the calibration function can be inverted properly, both functions should essentially describe the same thing. Problems may occur when inversion is difficult and rough approximations are used. But, as long as reasonable approximations are used, the difference should not be significant in comparison with the other uncertainties of the calibration. It is possible and necessary to make sure that the differences in the different approaches remain smaller than the measurement uncertainties.

The lesson from pure mathematics is to choose a set of regression parameters that shows little or no correlation (covariance), and to check carefully what you are doing.

4   Regression and Uncertainties of Calibration Parameters

In the following, we will use the approach of establishing the analytical function directly from the calibration data, rather than going through a tedious inversion of the calibration function.

Weighted Regression

The process of fitting the calibration function to the calibration data, *i.e.* of estimating the values of the parameters in the calibration function, is called regression. Though there are other ways to do it, such as the so-called robust fitting,[11] normally the parameters are estimated by the method of least squares. This method determines the most likely values for a set of fitting parameters, by assuming that errors in the data are normally distributed and that these errors are present only in the y-axis. In GDOES, the latter is not the case, as often there are significant errors in both axes. There are more sophisticated ways of handling regression with errors in both axes but these methods are rarely used in spectrochemistry. Instead, the errors in the x-axis are reflected into the y-axis when estimating the weighting function. This will be discussed later, in 'Combined uncertainties'.

The function χ^2 is the sum of the squares of the differences between the measured data points and the calculated points, divided by the associated variance:

$$\chi^2 \equiv \sum_{i=1}^{N} \left(\frac{y_i - y(x_i, a_1 \cdots a_m)}{\sigma_i} \right)^2 \qquad (4.11)$$

where σ_i is the standard deviation of the measurement y_i. In most real cases, σ_i is replaced by an unbiased estimate s_i. A weighted regression is the most appropriate for calculating calibration functions in GDOES, as the standard deviations usually cannot be assumed to be the same for all data points.

The method of least squares adjusts the parameter values to minimise χ^2. The minimum for a particular parameter is determined by taking the partial differentiation of χ^2 with respect to the parameter. If the calibration function is a linear combination of parameters, then the minimisation can be solved exactly and simultaneously for all parameters.

Note that when we refer to linear or non-linear regression, we are referring to linearity or non-linearity in the fitting parameters. Hence, a second-order polynomial ($c = a_0 + a_1 I + a_2 I^2$), for example, is linear in its three fitting parameters, a_0, a_1, a_2, and often non-linear equations can be rewritten so that the fitting parameters are linear, *e.g.* the non-linear calibration equation $(c = a_0(I - I_0)^{a_1})$ can be remodelled as a linear combination of a different set of parameters $(\ln(c) = a_0 + a_1(I - I_0))$.

If the calibration function contains non-linear components, then the minimisation can be done by progressive approximations until suitably close to a minimum. As there may be more than one minimum in a non-linear system, it is important to have good starting values for the parameters so that the minimisation process finds the desired minimum. For simplicity, we will concentrate here on linear least square fit procedures. The application to non-linear least square fits is left to the interested reader.

Because of errors in the measured data, the calibration function will not match the data perfectly, *i.e.* $\chi^2 > 0$, and there is some uncertainty in the fitted parameters. The smaller the value of χ^2, the lower the uncertainty. For a reasonable fit, on average or for large N, $\chi^2 \approx \nu$, where ν is the degrees of freedom, *i.e.* the squared deviations from the model equal the squared standard deviation in the data.[11] If χ^2 is close to ν, the model reproduces the measurement data within the limits of experimental uncertainty. If, on the other hand, χ^2 is far from ν, this may indicate that either the data are not representative (*e.g.* not enough points, outliers, *etc.*) or the uncertainties in the data have been grossly over or underestimated.

In linear regression, the general regression equation is of the form:

$$y(x) = \sum_{i=1}^{M} a_i X_i(x) \tag{4.12}$$

where a_i are the fit parameters and X_i are some functions of x. A common example is the polynomial, where $X_1 = 1$, $X_2 = x$, $X_3 = x^2$, ... Each data point is assumed to be an approximate solution to this general equation, leading to a matrix of such equations expressed as:

$$\mathbf{Y} = \mathbf{X} \cdot \mathbf{a} \tag{4.13}$$

which has the solution:[12,15]

$$\mathbf{a} = (\mathbf{X}^T \mathbf{V}^{-1} \mathbf{X})^{-1} \mathbf{X}^T \mathbf{V}^{-1} \mathbf{Y} = \mathbf{C} \cdot \mathbf{X}^T \mathbf{V}^{-1} \mathbf{Y} \tag{4.14}$$

where $\mathbf{V}$ is the 'variance matrix' (whose diagonal elements are the variances for each data point; off-diagonal elements are zero, *i.e.* we assume the separate measurements are uncorrelated). Its dimension is $N \times N$, N being the number of data points. Its inverse $\mathbf{V}^{-1}$ is called the weight matrix (whose diagonal elements are the inverse of the variances; again, off-diagonal elements are zero), and $\mathbf{C} = (\mathbf{X}^T \mathbf{V}^{-1} \mathbf{X})^{-1}$ is the covariance matrix. Its dimension is $M \times M$, M being the number of fit parameters.

The uncertainties, or estimation of the standard deviations, of the parameters a_i, can be calculated directly from the diagonal elements of the covariance matrix $\mathbf{C}$:

$$s^2(a_j) = C_{jj} \tag{4.15}$$

The off-diagonal elements give the covariance between the pairs of parameters. Detailed discussion and proof of this equation are beyond the scope of this work.

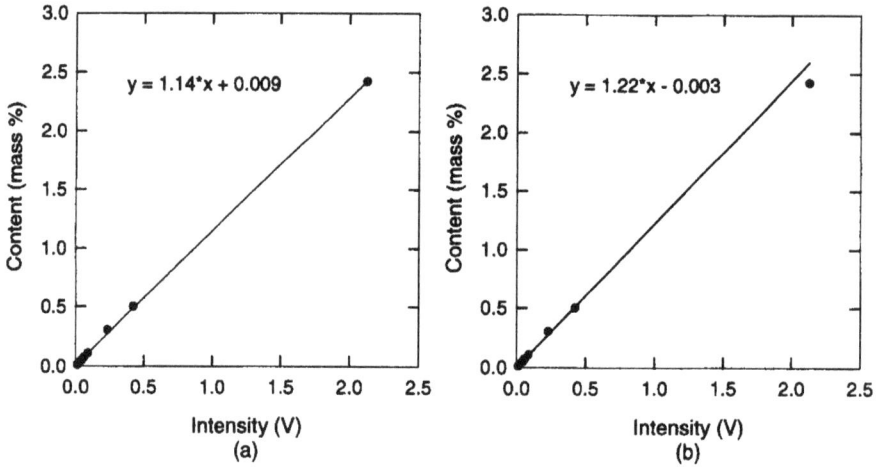

Figure 4.6   *Calibration for Cu 327 in cast iron: (a) unweighted and (b) weighted*

The interested reader may consult standard works on data fitting, such as those listed in Further Reading at the end of this book.

Non-weighted Regression

In some cases, when, for example, the uncertainties of the measured data are not available or the range of x and y values is small, non-weighted regression may be used. The uncertainties associated with the data points are assumed to be constant, independent of the measurement point. This assumption may hold when only small composition ranges are covered by the calibration. All uncertainties are generally set to unity, and:

$$\chi^2 = \sum (y_i - y(x_i, a_1 \cdots a_m))^2 \tag{4.16}$$

Once the regression calculation is performed, the residual χ^2 can be used to estimate the uncertainties of the single data points:

$$\sigma(y_i) = \sqrt{\frac{\chi^2}{n - p}} = \sqrt{\frac{\chi^2}{\nu}} \tag{4.17}$$

where n is the number of data points, p is the number of regression parameters and ν is the number of degrees of freedom. The uncertainty of the fitted parameters can then be estimated by:

$$\sigma^2(a_j) = C_{jj}\sigma^2(y) \tag{4.18}$$

However, this assumes a good fit and therefore does not allow an independent test of the quality of a regression model.

Figures 4.6 and 4.7 are displayed to show the importance of using appropriate weights in the calibration procedure. The linear plots in Figure 4.6 suggest to the eye

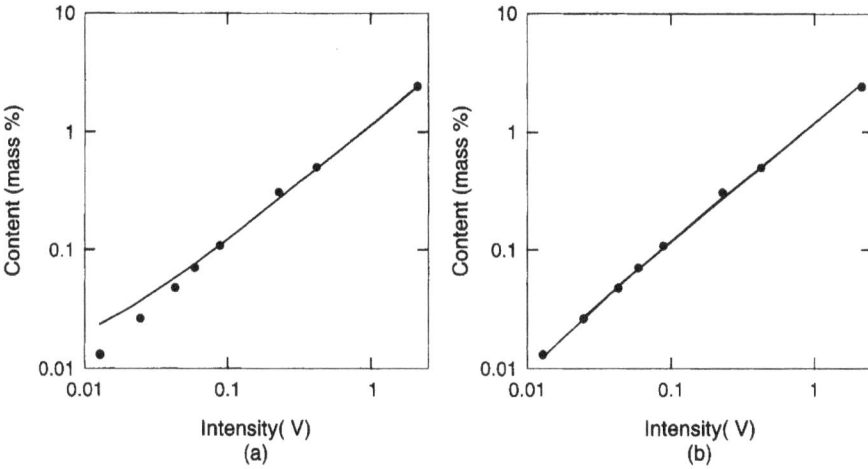

Figure 4.7 *Calibration for Cu 327 in cast iron: (a) unweighted and (b) weighted, each shown as log–log plots*

that the unweighted regression looks better. But the log–log plots in Figure 4.7 show clearly the problem of unweighted fit, particularly for low compositions.

Uncertainty Propagation

In calibration, the dependent variable, *e.g.* composition, is determined as some function f of measurements and parameters related to intensity. The uncertainty in the composition will then depend on the combination of the uncertainties of the different terms in this function. Following the recommendations given in the 'Fundamental standards Guide to the expression of uncertainty',[13] we can write the resulting variance in composition as:

$$\sigma^2(c) = \sum_{i=1}^{N} \sum_{j=1}^{N} \frac{\delta f}{\delta x_i} \frac{\delta f}{\delta x_j} u(x_i, x_j) \tag{4.19}$$

where x represents the parameters related to intensity, or, as we will see in this chapter, x could represent the calibration parameters, and $u(x_i, x_j)$ are the covariances between x_i and x_j.

When the parameters are independent, then:

$$u(x_i, x_j) = \sigma^2(x_i) \quad \text{when} \quad i = j$$
$$= 0 \quad \text{when} \quad i \neq j \tag{4.20}$$

and:

$$\sigma^2(c) = \sum_{i=1}^{N} \left(\frac{\delta f}{\delta x_i} \right)^2 \sigma^2(x_i) \tag{4.21}$$

To illustrate how Equation 4.19 works, consider some simple cases, where c is a function of two independent variables x_1 and x_2, *i.e.* we will use the simpler Equation 4.21:

1. For a simple sum or difference:

$$c = x_1 \pm x_2 \tag{4.22}$$

   we find the well-known formula for the uncertainties:

$$\sigma^2(c) = \sigma^2(x_1) + \sigma^2(x_2) \tag{4.23}$$

2. For products or ratios:

$$c = x_1 \times x_2 \text{ or } c = x_1/x_2 \tag{4.24}$$

   we find:

$$\sigma^2(c) = x_2^2\sigma^2(x_1) + x_1^2\sigma^2(x_2) \text{ or } \sigma^2(c) = \frac{1}{x_2^2}\sigma^2(x_1) + \frac{x_1^2}{x_2^4}\sigma^2(x_2) \tag{4.25}$$

   which can be expressed in the more familiar relative uncertainties:

$$\frac{\sigma^2(c)}{c^2} = \frac{\sigma^2(x_1)}{x_1^2} + \frac{\sigma^2(x_2)}{x_2^2} \tag{4.26}$$

Uncertainty of the Chemical Composition of CRMs

For many modern CRMs, the uncertainty of the composition is specified together with the composition. For many others, in particular older ones, they are not specified. In the latter case, it is wise to assume an uncertainty of the composition based on experience. Relative standard deviations in uncertainty of the composition of 1–5% are still optimistic for some CRMs. The uncertainties in the certified values should certainly not be neglected, *i.e.* they should not be set to zero.

When specific uncertainties are not available for individual samples, or to save time, a global approach is often used. The absolute uncertainty at trace levels is generally small, perhaps a few ppm, and tends to increase with composition. The uncertainty is therefore often assumed simply to be proportional to composition. The square of this uncertainty is then added automatically. The consequence is that data points at low composition will have greater weight in the regression, because of their lower uncertainty, than points at higher composition. This will compel the regression curve to pass closer to low composition points than to higher composition points. To the eye, this may not look the 'best' curve, as we saw in Figure 4.6, since some high points may be far from the curve, but analytically, if the estimated uncertainties and weights are reliable, then the curve shown is the best available.

Unlike the absolute uncertainty, the relative uncertainty is generally very high at trace levels, decreasing as the composition increases. The relative uncertainty in N content in nearly pure iron, for example, could be around 100%, while the relative uncertainty for Fe in nearly pure iron would be very much less than 1%.

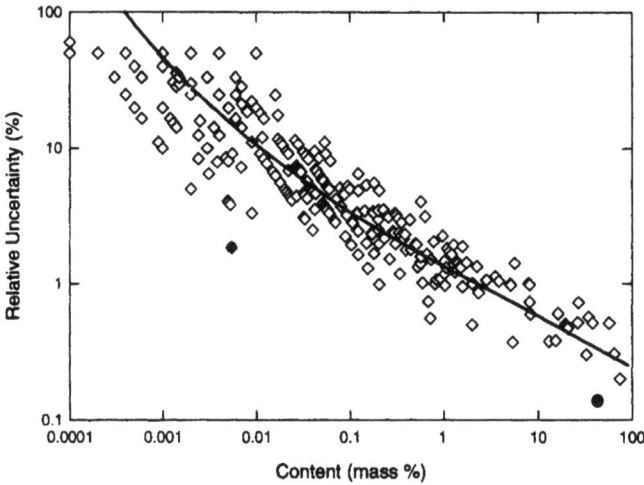

Figure 4.8 *Relative uncertainty of certified contents as a function of content (data from Weiss[14])*

Weiss has determined the relative uncertainty in content in a variety of reference materials as a function of content.[14] His data are shown in Figure 4.8 along with a line of best fit given by:

$$\log\left(\frac{\Delta c}{c}\%\right) = A + B\log(c\,\%) + C[\log(c\,\%)]^2 + D[\log(c\,\%)]^3 \quad (4.27)$$

where $A = 0.1282$, $B = -0.3735$, $C = 0.0194$ and $D = -0.0091$.

The original Weiss data do not include many points at high contents. So, we have decided to include some additional points using data available from the brass CRMs included in our brass calibration. When the expanded data are plotted as absolute uncertainty (rather than relative uncertainty) versus content, we obtain the intuitive result shown in Figure 4.9. It shows that uncertainties are low (in absolute terms) at low and high contents and peak at about 50% content. The fitted curve in Figure 4.9 is:

$$\Delta c\,\% = A' + B'(c\,\%) + C'(c\,\%)^2 \quad (4.28)$$

where $A' = 0.0030$, $B' = 0.0066$ and $C' = -0.000066$. This equation seems to show the expected trend at high contents but does not fit the data as well as Equation 4.27 at low contents. Therefore, in later calculations, where uncertainties were not available on certificates, Equation 4.27 was used for contents $\leq 10\%$ and Equation 4.28 for contents $> 10\%$.

Implicit in Equation 4.28 is the assumption that the content of the major element, approaching 100%, is calculated by difference, *i.e.* it is determined by subtraction of all measured impurities from 100%. This approach is often adopted for certified reference materials. The accuracy suffers if some of the impurities present in the sample are not included, especially gaseous impurities (*e.g.* O, H, N, *etc.*). If the

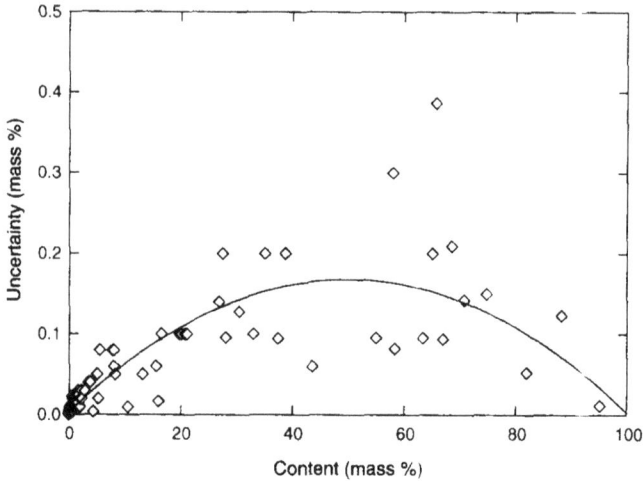

Figure 4.9   *Uncertainty of the certified composition as a function of composition*

major element is determined directly from calibration curves, then the trend might not bend as shown in Figure 4.9.

The certified compositions and associated uncertainties of CRMs are established using different analysis techniques in different laboratories, analysing different samples of the material. The given value and its uncertainty therefore represent the most likely average value of the composition and its associated uncertainty over larger parts of the material. In GDOES, we analyse only a few milligrams of the material. If the CRM is not perfectly homogeneous at this milligram level, we will analyse compositions with a different variance from those specified by the CRM manufacturer. If ε is the difference between the measured value and the certified value, the mean value of ε hopefully being zero, the measured composition then is:

$$c_a = \bar{c} + \varepsilon \tag{4.29}$$

This lack of homogeneity will show up in variations of the measured intensity together with other sources of variations. In this sense and only in this context can we justify ignoring the covariance term discussed below.

Combined Uncertainties

It is often assumed that the only source of uncertainty in calibration lies in the fluctuations of measured intensities. But, in general, uncertainties for each calibration point may arise both from fluctuations in the intensity and from uncertainties in the certified composition.

Consider five common carbon steel CRMs: BAS Nos. 456/1 to 560/1 (revised April 1988). The mean RSDs in composition for the certified elements in these samples vary from 2.0% for carbon to 10.8% for lead, with an average RSD for all elements of 6.1%.

If 15 such standard samples were used in the calibration, then the contribution to the standard uncertainty of estimate (defined later) would decrease on average to about $6.1\%/\sqrt{15-2} \approx 1.7\%$. So, a 6.1% RSD in each sample composition would produce a standard uncertainty of estimate comparable to or greater than typical uncertainties in GDOES intensities.

When there are uncertainties in both axes, the overall standard deviation in the dependent variable (y-axis) can be approximated as the square root of a linear combination of the variances due to each axis:

$$\sigma_i^2 = \sigma^2(y_i) + \left(\frac{\delta y}{\delta x_i}\sigma(x_i)\right)^2 \tag{4.30}$$

i.e. the overall variance can be approximated by the sum of individual variances.

It is important to realise that we are summing variances, not standard deviations. Because of the squared terms in Equation 4.30, if the uncertainties are about the same size, then the combined uncertainty is not the double of one of them but about $\sqrt{2} \approx 1.4$ larger. And if one is significantly larger than the other, then the combined uncertainty is close to the larger one. This means one large source of uncertainty will dominate and other sources of uncertainty become insignificant in comparison.

Let us take a more careful look into the estimation of the combined uncertainty in our dependent variable, the composition. It is a nice exercise in dealing with theory, reality, and uncertainties and approximations. In deriving Equation 4.30 by uncertainty propagation, we have ignored the covariance, *i.e.* the correlation between intensity and composition. Considering that here we are dealing with the calibration of a GD spectrometer, *i.e.* with establishing the correlation between intensities and composition, this omission seems a bit simplistic. Variations in the composition will obviously lead to variations in the measured intensities. If all variations in intensities were due to variations (uncertainties) in the composition, using Equation 4.30 we would have counted the effect twice. So how can we ignore the covariance?

Summarising the different sources of uncertainties, we can estimate the total variance of the dependent variable, composition, as a sum of three components:

$$\sigma^2(c) = \sigma_{CRM}^2(\bar{c}) + \sigma_{DL}^2 + \left(c\frac{\sigma(\bar{I})}{\bar{I}}\right)^2 \tag{4.31}$$

where c is the composition and $\bar{I}$ is the measured average intensity, $\sigma(\bar{I})$ is its standard deviation, $\sigma_{CRM}(\bar{c})$ is the standard deviation of the certified composition, and σ_{DL} is a constant term, representing the minimum uncertainty in the determination of a composition for the spectral line used. It should be roughly the detection limit, and is related directly to the intercept in Equation 3.5.

As examples, the combined uncertainties for the calibration of C 156 in low alloy steel are shown in Table 4.2. Those for the calibration of Al 396 in brass are shown in Table 4.3. The calibration curves are shown later in Chapter 5.

Table 4.2 *Uncertainties and weights used in regression for C 156 in low alloy steel*

Sample	Intensity	SD intensity	Content	SD content	SD combined	Weight
NBS 1761	1.354 7	0.000 83	1.03	0.014	0.014	5 400
NBS 1762	0.462 96	0.000 47	0.337	0.0068	0.0069	21 000
NBS 1763	0.281 3	0.001 9	0.203	0.0050	0.0053	36 000
NBS 1764	0.824 1	0.001 1	0.592	0.010	0.010	9 900
NBS 1765	0.018 47	0.000 39	0.006	0.0010	0.0014	520 000
NBS 1766	0.027 9	0.000 35	0.015	0.0012	0.0015	420 000
NBS 1767	0.083 35	0.000 58	0.052	0.0025	0.0027	140 000

Table 4.3 *Uncertainties and weights used in regression for Al 396 ratioed to Cu 225 in brass*

Sample	Relative intensity	SD relative intensity	Relative content	SD relative content	SD combined	Weight
B21	0.1617	0.0039	0.12	0.016	0.016	3 800
B22	0.286	0.016	0.24	0.012	0.018	3 100
B23	0.0965	0.0040	0.08	0.008	0.0087	13 000
B24	0.0073	0.0004	0.007	0.0022	0.0022	200 000
L2	0.7802	0.0038	0.49	0.008	0.0082	15 000
L3-182	1.458	0.017	0.91	0.011	0.016	4 200
L4	0.1650	0.0019	0.102	0.0005	0.0013	610 000
L5	0.3363	0.0057	0.200	0.0047	0.0058	30 000
L7	0.5615	0.0058	0.308	0.0036	0.0048	44 000
LH10	4.327	0.060	2.66	0.010	0.038	680
LH11	0.719	0.014	0.46	0.005	0.010	9 400
MNB1	0.6852	0.0098	0.48	0.007	0.0098	10 000
MNB2	0.3885	0.0080	0.26	0.006	0.0076	17 000
MNB3	1.1469	0.0061	0.75	0.023	0.023	1 800
MNB4	3.178	0.051	1.89	0.049	0.058	300
MNB5	5.690	0.036	3.27	0.047	0.051	380

5 Confidence Levels of Calibration Curves

Different statistical measures are used to express 'confidence levels'. It is fairly dangerous for the analyst to rely on only one type of measure to judge the quality of a calibration.

Most statistical measures are based on the assumption of normal distribution in the measurement uncertainty. This assumption is certainly not strictly valid for GDOES and most other OES techniques such as Spark and ICP. Non-normal distributions are often seen in measurements recorded over time, such as intensity measurements, as values at one time are often correlated with those immediately before and after. The result of this violation of 'normality' is that the estimated values for means and regression parameters remain unbiased but may not be the best available, and the predictions of significance and confidence intervals are compromised.[15]

Table 4.4 *Regression outcome for C 156 in steel (data from Table 4.2)*

$\bar{I}$	$\bar{c}$	r	r_{adj}	SUE	F
0.059	0.037	0.9993	0.9992	1.29	7160.3

Table 4.5 *Regression outcome for Al 396 in brass (data from Table 4.3)*

$\bar{I}$	$\bar{c}$	r	r_{adj}	SUE	F
0.193	0.119	0.9965	0.9926	3.28	2018.2

$\bar{I}$ is the mean intensity ratio to Cu 225 and $\bar{c}$ is the mean content relative to Cu.

Also, many of the formulae used are approximations for a large number of data points. Given that we constantly need to save time, we rarely use hundreds of calibration samples or repeat each measurement hundreds of times.

Once we accept that there is no single truth in calibration, and that we need to look at different parameters with some scepticism, it is quite possible to distinguish between good and bad calibrations. Some common measures are:

- correlation coefficient
- standard uncertainty of estimate
- F-parameter
- confidence limits

Given these measures, a typical process for optimising a regression involves selecting the calibration function that best matches the analytical task, *i.e.* selecting a set of fitting parameters so that the resulting regression gives a correlation coefficient close to 1, a low standard uncertainty of estimate (SUE), a high F-parameter, and acceptable confidence limits.

The results for the calibration of C 156 in steel are shown in Table 4.4. The results for the calibration of Al 396 in brass are shown in Table 4.5

When weighted regressions are used, or when uncertainties associated with the measurements are not constant, a weighted mean is used to express the different parameters. It de-correlates the regression parameters for a straight line. A weighted mean is no longer simply the sum divided by the number of points, but is given by:

$$\bar{x} = \frac{\sum (w_i x_i)}{\sum (w_i)} \tag{4.32}$$

The weight used should be the inverse of the variance or the experimental estimation of the variance, *i.e.* the inverse square of the standard deviation. The weights used for the calibration of C 156 in steel and Al 396 in brass are shown in Tables 4.2 and 4.3 and the weighted mean intensity and content are shown in Tables 4.4 and 4.5 respectively.

Correlation Coefficient

The correlation coefficient r, or its square, r^2 (also called the 'coefficient of determination'), is a measure of how well a regression function can be fitted to the data set. It ratios the variation about the mean, due to the regression, to the variation about the mean, due to the measurements. It therefore estimates the amount of variation accounted for by the regression.

It is usually calculated in one of the two ways: a generalised correlation coefficient

$$r = \left\{ \frac{\sum [w_i(\tilde{y}_i - \bar{y})^2]}{\sum [w_i(y_i - \bar{y})^2]} \right\}^{1/2} \tag{4.33}$$

or a linear correlation coefficient:

$$r = \frac{\sum [w_i(x_i - \bar{x})(y_i - \bar{y})]}{\left\{ \sum [w_i(x_i - \bar{x})^2] \sum [w_i(y_i - \bar{y})^2] \right\}^{1/2}} \tag{4.34}$$

They are equivalent only for a straight-line regression. The generalised version can be used for linear and non-linear functions because the function is included in the calculated values for $\tilde{y}_i$, while the linear version does not include $\tilde{y}_i$ explicitly, only the data values, and is only valid for straight lines.[16] This distinction is often not well known and it is important to determine which version is being used, because using the linear version for data clearly not fitting a straight line will give a low value for r even when there is obviously a high level of correlation.

In GDOES, intensities (or intensity ratios) often vary nearly linearly with composition for many materials, and r or r^2 values often approach 1. It is therefore not a very sensitive measure. It also depends strongly on the number of points in the regression and so care is needed while comparing values between different calibrations. It is a poor statistic and recommended only as a general first indicator.

When adding terms to the regression equation, such as higher orders in a polynomial, or corrections for spectral interference, *etc.*, r will tend to increase as new terms are added. It is therefore difficult to use r to determine which parameters should be included. For this reason, an adjusted r^2 is sometimes used, given by:[15]

$$r_{\text{adj}}^2 = 1 - \frac{(1 - r^2)(n - 1)}{(n - p)} \tag{4.35}$$

where p is the number of parameters (terms). r_{adj}^2 will approach 1 as important new terms are added but will stabilise at some maximum value when all important terms are included and then remain little changed when unimportant terms are added. The normal procedure is to include only those terms that make a significant change to r_{adj}^2.

Standard Uncertainty of Estimate

The standard uncertainty of estimate, $s_{y/x}$ (variously called the 'standard error of estimate', or, when squared, the 'residual mean square', the 'residual variance', or the 'variance of y given x'), is a measure of how well the standard deviation of the scatter of the calibration points about the calibration function matches the uncertainties used in the weights.

The standard uncertainty of estimate is defined as:

$$s_{y/x}^2 = \frac{\chi^2}{\nu} \tag{4.36}$$

where ν is the number of degrees of freedom, *i.e.* the number of data points minus the number of parameters. For a reasonably good fit, we expect the value χ^2/ν to approach unity, and $s_{y/x}^2 \approx 1$. In this case, the scatter around the regression function equals the experimental uncertainty, anything different indicates a problem that needs to be investigated (see discussion on χ^2).

The standard uncertainty of estimate is used to estimate the confidence limits of the calibration and, in particular, the standard deviation at the mean (centroid) of the calibration curve:

$$\sigma(\bar{c}) = \frac{s_{y/x}}{\sqrt{\sum(w_i)}} \tag{4.37}$$

When $\sigma(\bar{c})$ is divided by $\bar{c}$, it becomes a relative standard deviation of the mean, $r(\bar{c})$. Typical values of $r(\bar{c})$ in GDOES are 1–5%. The expected trend using Equation 4.37 for the calibration of Al 396 in brass is shown in Figure 4.2.

The standard deviation at the mean is a crucial measure of the success of the regression as this is the lowest standard deviation along the calibration curve. From Equation 4.37, provided $s_{y/x}^2 \approx 1$, the standard deviation of the mean can only be reduced by increasing the sum of the weights, which in turn can only be increased by reducing the uncertainties of the data points or by adding data points with low uncertainties.

When adding terms to the regression equation, $s_{y/x}$ will tend to decrease as important new terms are added but decrease only slightly or remain little changed when unimportant terms are added. The normal procedure is to include only those terms that make a significant change to $s_{y/x}$.

Though it is not commonly done, when the number of data points is less and the operator has confidence in the uncertainties in the data and the goodness of the regression model, then more reliable confidence limits may result from assuming $s_{y/x}^2 = 1$ rather than by using an unreliable value determined from Equation 4.36. This also includes the possibility of doing a calibration with $\nu = 0$, which is not possible using Equation 4.36.

F-parameter

The F-parameter is often quoted in regression and statistics but perhaps not widely used for the optimisation of calibration curves. It is the ratio of the mean square due to regression MS_{reg} and the mean square due to the residual variance, *i.e.* the square of standard uncertainty of estimate, $s_{y/x}^2$, and follows an F-distribution. It can be used to estimate the likelihood that the regression function is better than an accidental fit to the data. So:

$$F = \frac{MS_{reg}}{s_{y/x}^2} \tag{4.38}$$

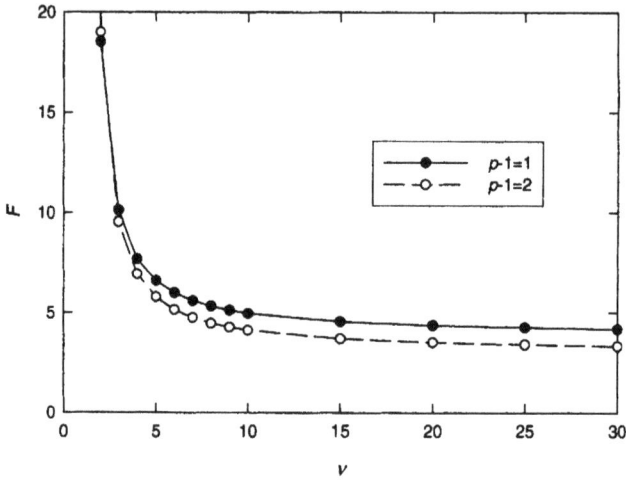

Figure 4.10 *Percentile values for the F(p − 1,v,0.95) distribution*

where the mean square due to regression is given by:

$$MS_{reg} = \frac{(\hat{c}_i - c_i)^2}{v} \tag{4.39}$$

where $\hat{c}_i$ is the calculated value from the regression function. For a straight-line calibration:

$$MS_{reg} = s_{y/x}^2 + b_1^2 \sum (I_i - \bar{I})^2 \tag{4.40}$$

and

$$F = 1 + \frac{b_1^2 \sum (I_i - \bar{I})^2}{s_{y/x}^2} \tag{4.41}$$

The hypothesis that the regression is a good fit is tested at 95% confidence levels by checking that the value of the F-parameter is larger than the tabulated values of the $F(p − 1,v,0.95)$ distribution, where p is the number of fitting parameters, and v is the number of degrees of freedom.[12] The variation of F with v for two different values of $p − 1$ is shown in Figure 4.10. The limiting value for $F(1,\infty,0.95)$ is 3.84.[17]

It is worth noting that a good fit means that the regression has succeeded in fitting the data but it does not necessarily mean a 'worthwhile' fit, in the sense that, for analysis, we need the working range of the regression to be much greater than the uncertainties of the regression. This implies that the F-parameter should be much higher than a simple pass or fail on the F-test, perhaps 5–20 times the F-distribution value. So, typically, F should be 50 or more.

Since $s_{y/x}^2 \sim 1$ in the ideal case, F will tend to increase when samples are added to the regression whose intensities are far from the mean (either high or low), for example, by forming 'clouds' of samples at either extreme of the regression curve or by increasing the range of the regression. Note that F will only improve with an extended range if the regression model remains valid over the extended range since $s^2_{y/x}$ will increase as the model becomes less valid.

Confidence Limits

The uncertainty at any point on the calibration curve will depend on the combined uncertainties in the fitting parameters. This will be discussed in more detail later, but for the moment let us consider a simple linear calibration function. For convenience, this function can be represented as a function about the mean:

$$c = \bar{c} + b_1(I - \bar{I}) \tag{4.42}$$

Using the average values $\bar{c}$ and $\bar{I}$ makes sure the regression parameters $\bar{c}$ and b_1, and the uncertainty at any point is:

$$\sigma^2(c) = \sigma^2(\bar{c}) + \sigma^2(b_1)(I - \bar{I})^2 \tag{4.43}$$

It is then possible to show that:

$$\sigma^2(\bar{c}) = \frac{s^2_{y/x}}{\sum(w_i)} \tag{4.44}$$

and, for a straight line:

$$\sigma^2(b_1) = \frac{s^2_{y/x}}{\left[\sum w_i(I_i - \bar{I})^2\right]^{1/2}} \tag{4.45}$$

It is worth noting that the uncertainty in the slope can be reduced by including data points for which $(I_i - \bar{I})^2$ is large, *i.e.* including points at the extremes of the calibration range (or by adding calibration samples to increase the calibration range, provided the calibration function is still valid over the extended range).

Combining these equations, we obtain:

$$\sigma^2(c) = s^2_{y/x}\left\{\frac{1}{\sum w_i} + \frac{(I - \bar{I})^2}{\sum w_i(I_i - \bar{I})^2}\right\} \tag{4.46}$$

Finally to determine the uncertainty at a particular confidence level, we must multiply by the appropriate value of the Student's t parameter:

$$\Delta c = t\sigma(c) = ts_{y/x}\left\{\frac{1}{\sum w_i} + \frac{(I - \bar{I})^2}{\sum w_i(I_i - \bar{I})^2}\right\}^{1/2} \tag{4.47}$$

The confidence limit is a minimum at the mean intensity ($I = \bar{I}$) and diverges at either end of the analytical curve ($I \ll \bar{I}$ or $I \gg \bar{I}$), hence the most reliable analyses will be towards the mean of the calibration range (see Figure 4.11). The divergence is mainly due to the uncertainty in slope. It can also be seen that the confidence limit (generally) improves with an increased number of calibration samples, because both $\sigma(c)$ and t decrease.

In weighted regression, because the mean is generally shifted far from the centre of the curve, the confidence limits usually look quite different from those shown in

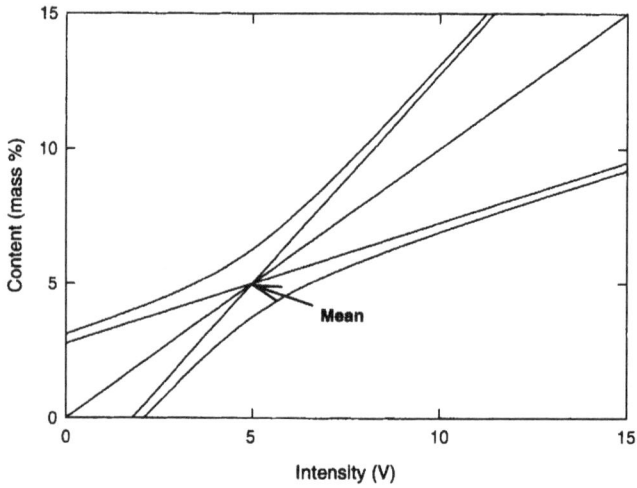

Figure 4.11   *Idealised regression line showing upper and lower confidence limits and asymptotic lines for the uncertainty in slope about the mean*

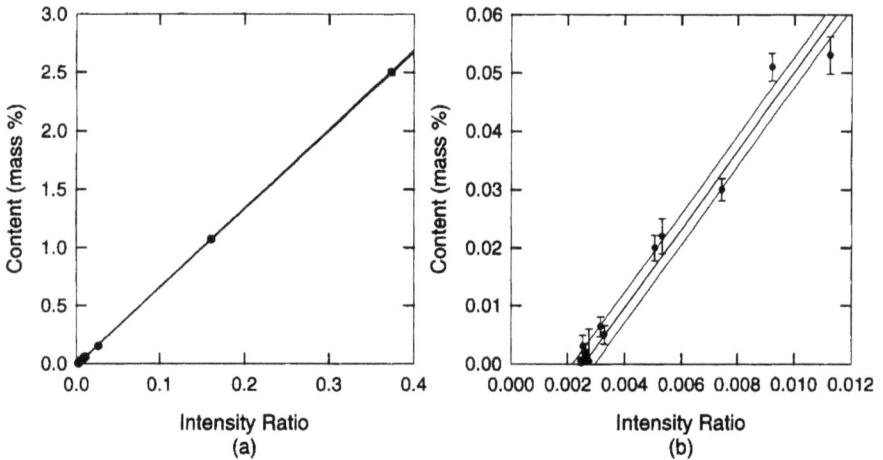

Figure 4.12   *Weighted calibration curve for Ni in Al–Si, showing combined uncertainties and 95% confidence levels, after omitting the outlier described in Section 7. In the full curve (left), the uncertainties and confidence levels are so small as to be difficult to see. The expanded curve (right) shows the region near the origin*

Figure 4.11. Also, when seeking accurate analysis, confidence limits should be close to the calibration curve. A real example is shown in Figure 4.12a and expanded near the origin in Figure 4.12b, where, in approaching the detection limit, the uncertainties are greatest.

In equally weighted regression, uncertainties in regression parameters tend to decrease as the square root of the number of samples or measurements, see for example

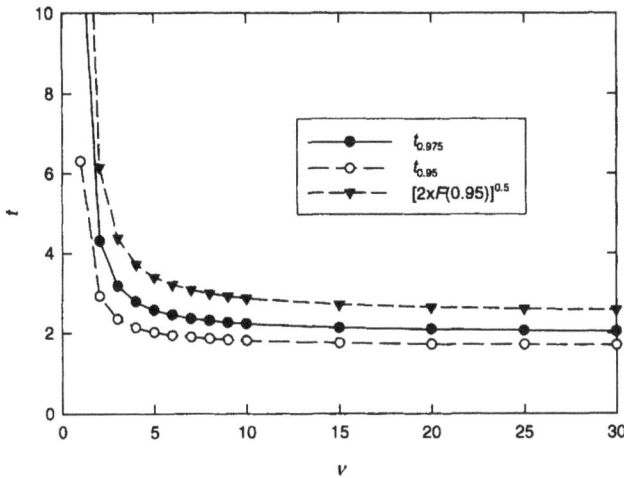

Figure 4.13 *Percentile values for Student's t-distribution for 90% ($t_{0.95}$) and 95% ($t_{0.975}$) confidence levels, where ν is the number of degrees of freedom, compared with the F-distribution parameter for a 95% confidence level*

Equation 4.17. In weighted regression, they tend to decrease as the square root of the sum of weights. So a sample or measurement with a small weight will have little effect, while a sample or measurement with a large weight could have a large effect (see Figure 4.2).

Student's *t*-Distribution

The Student's t parameter is defined as:

$$t \equiv \frac{\bar{x} - \mu}{s/\sqrt{\nu}}$$

where s is an unbiased estimate of the true standard deviation, σ. Student's t-distribution is the distribution of the Student's t parameter, and approaches a normal distribution as ν increases. In effect, it is a measure of the likelihood $s \to \sigma$.

The Student's t-parameter is therefore a measure of the uncertainty of the distribution of measurements. The greater the number of points, the more likely the actual distribution of measurements matches a normal distribution. As the number of points approaches infinity, the Student's t percentile (*i.e.* the integral of Student's t-distribution) approaches a lower limit. The variation of t with n for two different values of confidence ($t_{0.95}$ is 90%, $t_{0.975}$ is 95%) is shown in Figure 4.13. For 95% confidence ($t_{0.975}$), the lower limit for t is 1.96 (often approximated to 2).

Also shown for comparison is the F value at a 95% confidence level. In a linear calibration there are two varying parameters, so it would be preferable to use the F-distribution instead of Student's t-distribution, *i.e.* replace t in Equation 4.47 by $[2F(2, \nu, 0.95)]^{1/2}$.[18,19] The limit of this parameter for an infinite number of points is 2.45. The use of the F-distribution would therefore increase the estimated uncertainty. But to date, it is the Student's t parameter that is generally used in spectrochemistry.

6 Outliers

Outliers are data points that are so far from the regression line that they should be eliminated and the regression repeated. The further a point is from the mean intensity, the more effect it can have on the regression.

These data can arise from purely statistical means, *i.e.* a 1 in 100 chance that the noise is too far in one direction. With 20 points, for example, in a calibration, such a point will unduly shift the regression curve. Fortunately, such points should occur very rarely, perhaps one in five calibrations. More likely, such a point arises from a problem during the measurement, for example, a short circuit or air intake. Such points will generally have much lower intensities than expected. The measurement should be repeated to verify this and if the problem continues (*e.g.* the sample is porous), the point should be omitted from the regression. Occasionally, the intensity may be much higher than expected. This could be due to a noise spike in the detector or perhaps due to H in the sample. Normally, the software should alert you to the presence of noise spikes. If there is no noise spike, check the H signal from the sample. Occasionally, outliers are due to a wrong value of the composition or relative sputtering rate in the sample database. If this is the case, correct the error and, if necessary in the software, repeat the measurement.

Outliers can also result from inadequacies in the regression model. For example, if a bulk method is used without correction for sputtering rates and one sample has a different matrix, and consequently a different sputtering rate from the others, then it will appear as an outlier. The solution would be to exclude such matrices from the method or to include a sputtering rate correction.

A suspected outlier is shown in Figure 4.14 for Ni 341 in an Al–Si calibration. The intensity is higher than would be expected from the regression line if this point were excluded from the regression. The intensity SD for this point was a little high, possibly indicating a spike, so it would be worth repeating the measurement on this sample, to be sure, rather than just omitting it. Another possibility worth considering is that calibration samples GCH1 and 2174-2 are both low in Si while samples 433-01 and AS40-1 have high Si, so a relative (ratio) calibration method might be better (see Chapter 6, for more details), *i.e.* consider modifying the regression function.

A common way of detecting outliers is to plot residuals (*i.e.* deviations from the expected value), but when weighted regression is used the residuals should be divided by their standard deviations and standard uncertainty of estimate, *i.e.* to use weighted standardised residuals. If the uncertainties are normally distributed, most of these should vary between –2 and 2, with outliers typically being outside ±2.5.

As an example, consider Al 396 in our brass calibration shown previously in Figure 4.4. Visual inspection of Figure 4.4 suggests that the high point may be an outlier. This is supported by the absolute values of the residuals, shown in Figure 4.15a, but the weighted standardised residuals, shown in Figure 4.15b, suggest the variation of this point is normal, and that there are probably no outliers. In contrast, if we examined the suspected outlier in Figure 4.14, we would find that its weighted standardised residual was –3.7, suggesting it should be omitted.

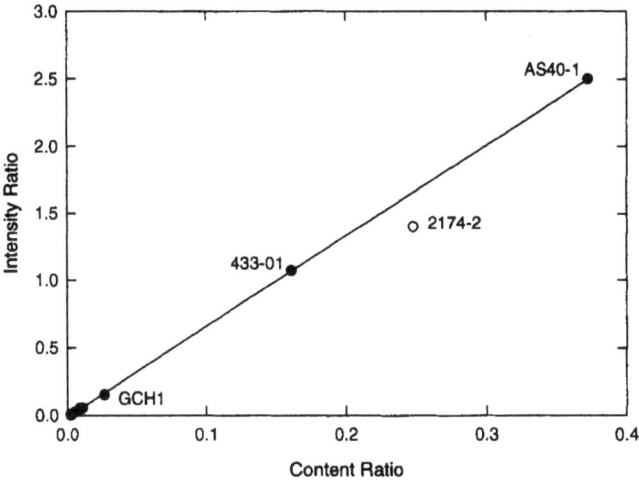

Figure 4.14 *Calibration curve for Ni in Al–Si, showing a suspected outlier (2174-2)*

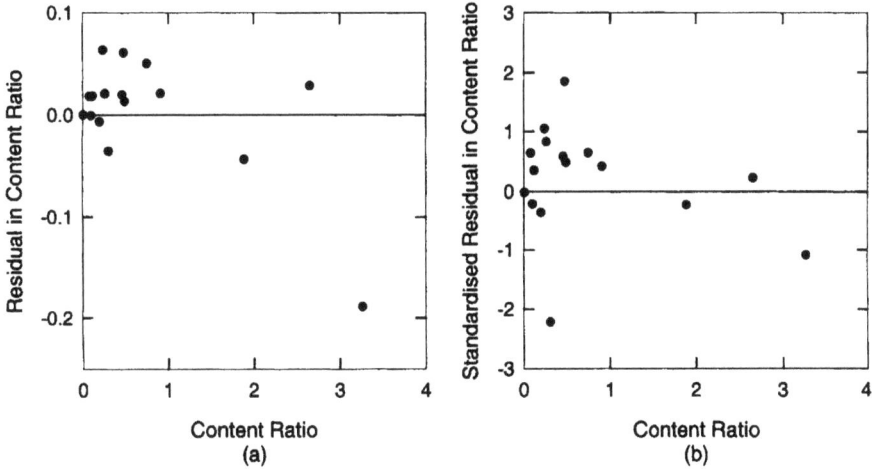

Figure 4.15 *Al 396 in brass: (a) absolute value of residuals and (b) absolute value of weighted standardised residuals versus Al content, for data shown in Figure 4.4b*

The greatest danger with outliers is that the operator will simply delete one or more data points, without valid reason, simply to make the regression look better. The danger is that these omitted points may contain valuable information, and by omitting them the operator will have a more restricted data set. Any subsequent analysis could then be adversely affected, since the samples being analysed might be similar to those omitted. The lesson is: unless you are very sure a point is not valid, do not omit it. Accept the greater uncertainty in the calibration or deal with the variation (by including more samples and suitable corrections, *e.g.* for spectral interference, *etc.*).

7 Optimisation of Calibration Curves

Once the calibration has been made, the resulting curves need to be optimised for each element. This optimisation involves: checking that all points for an element are acceptable (see outliers), the weights are suitable for each point, and selecting the best calibration function.

The process of selecting the best calibration function includes selecting the overall function (*e.g.* polynomial) and deciding which corrections (if any) should be included. It is a process that requires knowledge and experience to obtain the best results. Various examples will be shown in Chapter 5 for bulk analysis and in Chapter 6 for compositional depth profiling.

A key decision is whether the calibration curve is linear (straight line) or a higher order polynomial. Calibration curves are usually non-linear because of self-absorption. In GDOES, for significant self-absorption to be present in an emission line:

- The emission line must be a resonance or near-resonance atomic line, *i.e.* with a state I R or I r.
- The emission line must have a large self-absorption coefficient; see Appendix A.
- The element must have a high atom density in the plasma, *i.e.* a high content in the sample and a high sputtering rate.

This topic is discussed in more detail in Chapter 11.

Consider the brass calibration data in Table 4.1. Of the ten elemental lines, six do not have I R or I r states and so should be linear. Fe 371 has an I R state but has a relatively low self-absorption coefficient, and usually shows little effect on self-absorption. P 178 has only a low content in the samples, and so should be linear. This leaves only Al 396 and Ni 341, that are of concern. As shown in Figure 4.4, the Al 396 calibration appears linear; the Ni 341 line, though not shown, also had no significant improvement in second order. This is probably because of the relatively low contents of these elements in the samples used for calibration.

8 Validation of Calibration Curve

Once the calibration curve is established and the obtained precision is satisfactory it should be checked and validated, to make sure that no doubt about its applicability persists.

A number of certified reference materials covering the field of application of the calibration curves should be analysed. The certified compositions should be reproduced within the uncertainty of the calculated composition, including the standard deviation of the intensity measurement as well as the effect of the uncertainty on the calibration parameters. This will be discussed further in Chapter 8. When checking the analysis with the certified values, also consider the uncertainties in the certified values if these are given on the certificates or the generalised uncertainties are shown in Figures 4.8 and 4.9:

- Analysing CRMs that are already included in the calibration process will check and validate the internal consistency of the analysis process.

- Analysing CRMs not included in the calibration process will check and validate the external consistency of the analysis process.

If some CRMs were excluded from the calibration process simply because they did not fit, it is wise to re-analyse them to check whether their composition is now reproduced and the error was just some accidental bad measurement. If the discrepancy persists, we suggest looking for a different CRM of similar composition and checking whether its composition is reproduced. If the discrepancy still persists, there is a problem and samples of similar composition should not be analysed with this set of calibration curves.

References

1. Norme Internationale ISO 10012-1, Exigence d'assurance de la qualité des équipements de mesure, Partie 1, 1992.
2. ISO 17025, General requirements for the competence of testing and calibration laboratories, 2002.
3. ISO, 2002, www.iso.ch/iso/en/aboutiso/introduction/
4. ISO 14707-2000, Surface chemical analysis—glow discharge optical emission spectrometry (GDOES)—Introduction to use, 2000.
5. *EURACHEM/CITAC Guide Quantifying Uncertainty in Analytical Measurement*, 2nd Edn., EURACHEM/CITAC, 2000.
6. International Vocabulary of Basic and General Terms in Metrology, ISO, Geneva, 1993.
7. NIST Policy on Traceability, NIST, USA, 2002, www.nist.gov/traceability/
8. ISO International Vocabulary of Basic and General Terms in Metrology, 2nd Edn., 1993, definition 6.10.
9. International Union of Pure and Applied Chemistry, *Spectrochim. Acta*, 1978, **33B**, 247.
10. M. Abramowitz and I.A. Stegun, *Handbook of Mathematical Functions*, Applied Mathematics Series 55, National Bureau of Standards, Washington, DC, 1965, 16.
11. W.H. Press, S.A. Teukolsky, W.T. Vetterling and B.P. Flannery, *Numerical Recipes in Fortran*, Cambridge University Press, Cambridge, 1992, 655–7, 694.
12. N.R. Draper and H. Smith, *Applied Regression Analysis*, John Wiley & Sons, New York, 1981, 94, 109, 131.
13. Norme Européenne NF ENV 13005, X 07-020, Guide pour l'expression de l'incertidude de mesure, 1999.
14. Z. Weiss, *J. Anal. Atom. Spectrom.*, 2001, **16**, 1275.
15. J.O. Rawlings, S.G. Pantula and D.A. Dickey, *Applied Regression Analysis*, 2nd Edn., Springer, Berlin, 1998, 28–9, 222–3, 329, 414–7.
16. M.R. Spiegel, *Theory and Problems of Probability and Statistics*, Schaum's Outline Series, McGraw-Hill, New York, 1980, 263–4.
17. M.R. Spiegel and J. Liu, *Mathematical Handbook of Formulas and Tables*, Schaum's Outline Series, McGraw-Hill, New York, 1999, 267.
18. C. Daniel and F.S. Wood, *Fitting Equations to Data*, John Wiley & Sons, New York, 1999, 12.
19. D.L. Massart, B.G.M. Vandeginste, L.M.C. Buydens, S.De Jong, P.J. Lewi and J. Smeyers-Verbeke, *Handbook of Chemometrics and Qualimetrics: Part A*, Elsevier, Amsterdam, 1997, 195, 432–5.

CHAPTER 5
Calibration for Bulk Analysis

This chapter will concentrate on aspects of calibration specific to bulk analysis and will deal with different calibration types, internal references and mini-calibration.

1 Calibration Types

The general calibration function for GDOES, described later in Chapter 11, is:

$$c_i q_M/q_{Ref} = k_i R_i S_i I_i - b_i + \sum_j d_j I_j \qquad (5.1)$$

where c_i is the composition of element i, q_M/q_{Ref} is the relative sputtering rate of the matrix M, k_i is the instrument constant, R_i is relative inverse emission yield, S_i is relative inverse self-absorption coefficient, I_i is emission intensity of element i, b_i is background term, I_j is intensity of a spectral line from interfering element j and d_j is the relative size of the interference.

When it comes to bulk analysis, we normally only have to analyse one matrix at a time, so some simplifications can be made to Equation 5.1.

Nearly Pure, Single Element Materials

For nearly pure materials with a matrix formed by a single element, *e.g.* nearly pure iron, copper or aluminium, we can assume that the sputtering rate is constant from one sample to the next. Also, we can assume that there is no change in the plasma from one sample to the next, and hence no change in emission yield. The compositions of all the elements other than the matrix element will be small, so their intensities should show no significant self-absorption, though there may be a nearly constant amount of self-absorption in the resonance lines of the major element. Hence Equation 5.1 reduces to:

$$c_i = k_i I_i - b_i + \sum_j d_j I_j \qquad (5.2)$$

From Equation 5.2 we expect that all calibration curves for minor and trace elements will be linear.

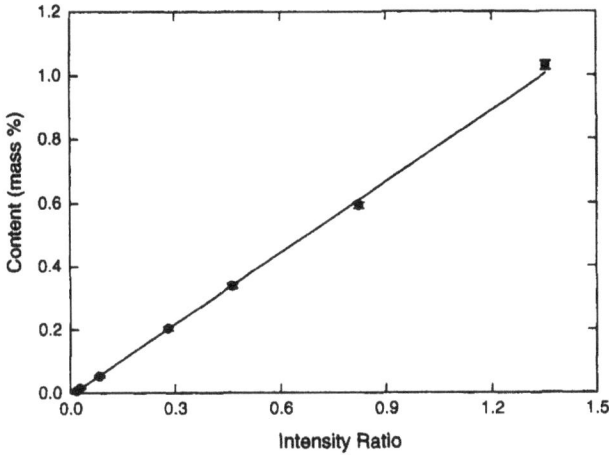

Figure 5.1 *Analytical calibration curve for C 156 in low-alloy steel, where C 156 intensity is ratioed to Fe 372 intensity*

The background term, in this mode, is called the background equivalent composition (BEC). By analogy, the background term in Equation 5.1 is also often called the BEC, though strictly speaking it is a sputtering rate adjusted BEC. Typical values for the BEC in GDOES range between 10 and 900 ppm.[1,2] As a rule of thumb, detection limits are roughly equal to BEC/30, *i.e.* typically between 0.3 and 30 ppm, depending on the element and the emission line.

To improve precision, it is common to measure the intensities relative to some internal reference, such as the intensity of the major element, or to an argon line or the total light from the GD source. This can reduce burn-to-burn variations. The calibration equation then becomes:

$$c_i = k_i I_i / I_{\text{Ref}} - b_i + \sum_j d_j I_j / I_{\text{Ref}} \qquad (5.3)$$

and the compositions c_i are plotted against the relative intensities I_i / I_{Ref} (see, *e.g.*, Figure 5.1). Though not yet common, in fact, it is quite sensible to use different reference lines for different analytical lines, based on their dependence on external perturbations, a technique frequently used in Spark analysis. With the introduction of solid-state detectors, this possibility will certainly find wider application.

This approach normally works well for nearly pure materials but the operator should be aware when the assumptions, especially about constant sputtering rates, are no longer valid. The sputtering rate is determined largely by the equilibrium surface composition during sputtering rather than by the bulk composition. This means that slowly sputtering elements are preferentially enriched on the surface and can alter the sputtering rate (and the plasma) much more than their bulk composition would suggest.

Consider, for example, an extreme case: small amounts of Al in zinc. Al sputters at only about 1/20 the rate of Zn, hence for low levels of Al in zinc, the surface Al will be about 20 times its bulk composition. So 0.5% Al, for example, would produce sputtered surface of about 10% Al and diminish the sputtering rate by about 10% compared with pure zinc. This effect would introduce curvature in the Al calibration curve, but would not introduce scatter, because it would maintain a good correlation between Al content and Al intensities. But it would increase the scatter in the calibration curves of other elements since their intensities are now correlated with the changes in sputtering rate, and therefore with Al content, rather than just with their own compositions.

Alloys

For alloys and other materials with more than one major element, *e.g.* brass, stainless steel and Al–Si alloys, we can restrict the range of compositions and the source operating conditions so that the relative emission yields are constant. Alternatively, we could introduce corrections for changes in emission yield, so that the corrected intensities have constant emission yield. If we then choose, as a reference, a strong emission line for one of the major elements, labelled R, which is not affected by self-absorption, then its intensity will be given by:

$$c_R q_M / q_{Ref} = k_R I_R \tag{5.4}$$

where the background signal and spectral interferences are negligible for this strong line. If we now divide Equation 5.2 by Equation 5.4, we have a calibration equation:

$$c_i / c_R = k_i I_i / I_R - b_i / I_R + \sum_j d_j I_j / I_R \tag{5.5}$$

The relative compositions c_i/c_R, are now plotted against the relative intensities I_i/I_R. See, for example, Figure 5.2. We will refer to this as the 'relative method', but it is also known as the 'ratio method', 'normalised method' or 'virtual method'. How this method is used in analysis will be described in Chapter 8. Note that we have not made any assumptions about sputtering rates, only about emission yields, hence this equation should work even if the sputtering rates change from sample to sample.

This approach normally works well for many materials but the operator should be aware of cases when the assumptions, especially about the reference line, are no longer valid. While it is usually possible to choose a reference line that has negligible background and spectral interferences, it are not always possible to select a line from those available on the instrument that are not subject to self-absorption.

Negative correlations in the compositions of the major elements are often present because their compositions add up to nearly 100%; so, when the composition of one element (*e.g.* Cu in brass) increases, the composition of another element (*e.g.* Zn in brass) decreases. Significant self-absorption in the reference line will therefore tend to bend the calibration curves for correlated elements and add scatter to those elements that are not correlated.

Samples with differing compositions of the reference element will have differing amounts of self-absorption in the reference line and will therefore be separated into different families in the calibration curve. An example is shown in Figure 5.3, for Al 396 ratioed to the extremely non-linear Cu 325 line. Compare this with the Al

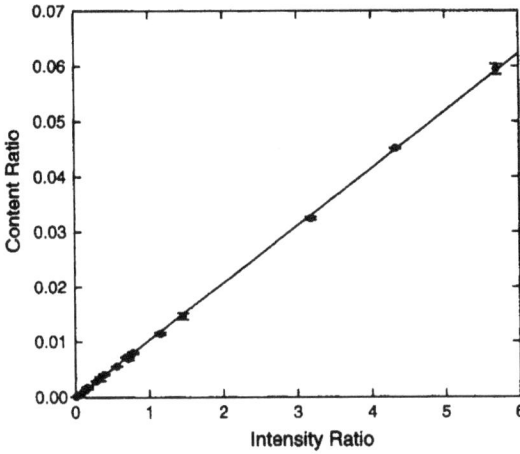

Figure 5.2 *Analytical calibration curve for Al 396 in brass, with intensity of Al 396 ratioed to Cu 225, and content of Al ratioed to content of Cu*

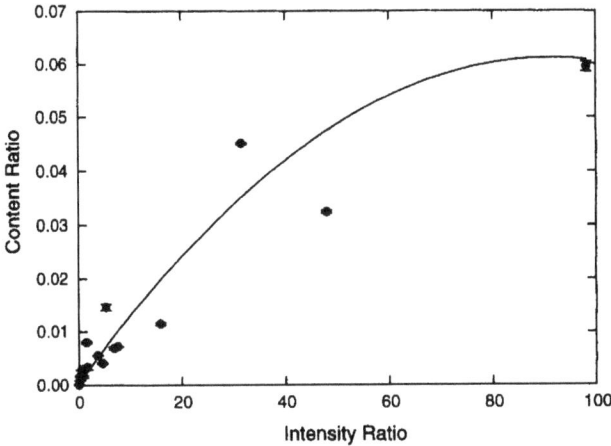

Figure 5.3 *Analytical calibration curve for Al 396 in brass, with intensity of Al 396 ratioed to the extremely non-linear Cu 327, and content of Al ratioed to content of Cu*

calibration in Figure 5.2, where Al 396 was ratioed to the linear Cu 225 line. In all cases of non-linear reference lines, the analytical results will be adversely affected. The given example was chosen to show how severe the problem can be for a 'text-book' case. In many other examples, reference lines showing less self-absorption and smaller changes in sputtering rates are used, and the effect on analytical results is less visible, but the results will still be adversely affected. Indeed, in less severe cases, non-linear reference lines may be more dangerous as their effect could be unnoticed, and later, poor analytical results might remain unexplained.

So, where possible, choose a linear reference line. This can be done by:

- checking that it is not a resonance or near-resonance atomic line, or by checking that its self-absorption coefficient is not too high. This information is available for many lines in Appendix A.
- choosing several lines of the same element (if available), to make optimal couples. The ability of CCD spectrometers and monochromators to provide a wide variety of possible lines could be an advantage in this context.

If a resonance reference line cannot be avoided, restrict the range of compositions considered in the calibration, and the calibration samples, to a single material type.

To take full advantage of this powerful analytical method, it is important to intelligently choose a couple of analytical line and reference. When looking at the depth profile of a homogeneous sample, intensity variations are often observed during the first minute(s). Some spectral lines will tend to increase, and others will tend to decrease, mainly due to surface preparation and residual gases in argon, especially hydrogen. In bulk analysis, the pre-burn is designed to minimise this variation, though it may not be completely removed. Employing the ratio of two lines, both increasing or both decreasing over time, for example, will lead to more stable results than choosing an analytical line and a reference line, one increasing and the other decreasing.

Global Method

The approach described later in Chapter 6 for compositional depth profiling can also, of course, be used for bulk analysis; indeed, it is a powerful means for analysing unknown samples and complex matrices, such as Ni–Co–Cr–Fe alloys. The calibration for a wide range of matrices is conducted identically to the procedure described in Chapter 6 for depth profiling, *i.e.* a calibration is made for $c_i q_M / q_{Ref}$ versus I_i.

The main advantage of this analysis mode is that it will work for many different materials. The main drawback is the large uncertainty in the calibration constants due to the uncertainties in the relative sputtering rates. This can be partly improved by using a large number of calibration samples. We expect the use of this analysis mode, unique to GDOES, to spread in future as reliable experimental data on sputtering rates become available and the understanding of multi-matrix calibration continues to improve.

For samples of completely unknown composition, the method can be used as a preparation step to give an approximate composition, followed by analysis with a more specific calibration.

2   Mini-Calibration (Type Standardisation)

Mini-calibration is a popular means for analysing a small number of samples. It is also known as 'type standardisation' or the 'nearby technique'. Although it lacks official sanction by standards organisations, it often gives the best results if handled carefully. Strictly, it is not really a calibration at all but an adjustment to an existing calibration. A known sample very similar to the samples to be analysed is chosen and analysed. It is essential that only those elements in the mini-calibration sample that are known with confidence, and are similar to the samples being analysed, be used in the mini-calibration. All other elements should be left unchanged or drift corrected in the usual way.

After running the mini-calibration sample, the calibration coefficients are adjusted automatically by additive or multiplicative factors in the software to give the correct result for this mini-calibration sample. Then, the other samples are analysed using these adjusted calibration coefficients.

When the analysis is finished, check that the analysed samples are indeed similar to the mini-calibration sample.

Mini-calibrations are especially interesting when the required accuracy for the measurement is particularly demanding. To translate the experience of many analysts, it is best to compare an unknown sample with a known sample very similar in composition. It can drastically improve the accuracy of the data, especially when the calibration curves are poor, *i.e.* when $s_{y/x}^2 \gg 1$.

Mini-calibration might find official sanction by standards organisation if it were associated with a proper treatment of uncertainties. In this case, however, we would notice that even though mini-calibration can improve the accuracy over a small range of compositions, it will not improve the precision of the measurements.

3 Sequence for Calibration for Bulk Analysis

The following steps are recommended for calibration for bulk analysis:

Create method
1. select elements
2. select appropriate spectral lines
3. select calibration, recalibration and sputter reference samples
4. check that composition ranges of calibration samples match analysis needs
5. select source conditions

Calibrate
1. prepare samples
2. make calibration
3. optimise regression
4. check drift-correction (recalibration) standards
5. validate calibration

References

1. R. Payling, D.G. Jones and S.G. Gower, *Surf. Interface Anal.*, 1993, **20**, 959.
2. Z. Weiss, *J. Anal. Atom. Spectrom.*, 2001, **16**, 1275.

CHAPTER 6

Calibration for Compositional Depth Profiling

The purpose of this chapter is to give advice on how best to calibrate a GDOES instrument for compositional depth profile (CDP) analysis. The theoretical background for CDP will be given later, in Chapter 11, as part of the theory of GDOES. But to help in understanding the requirements of calibration for CDP, we will begin with a brief outline of the basics of the quantification model.

Calibrating a GDOES instrument for CDP analysis is essentially the same as calibrating for bulk analysis, so comments on calibration in the preceding chapters are fully applicable to calibration for CDP, with two exceptions. For CDP, the calibration:

- is generally multi-matrix, *i.e.* it covers a range of different materials
- must contain information on sputtering rates

Calibrating for CDP means establishing a link between intensity and the number of atoms removed from the sample surface.

When the plasma is ignited, argon ions bombar the sample surface and sputtering removes atoms from the sample surface into the plasma. The sample surface is devoid of easily sputtered atoms and enriched with more difficult to sputter atoms. The effect is called preferential sputtering. The altered surface layer is about 2 nm thick,[1] so the theoretical depth resolution is about 2 nm. Despite preferential sputtering, the rate of atoms being removed from the surface is proportional to their composition in the sub-surface, *i.e.* stoichiometric (or steady-state) sputtering. This is a result of the conservation of mass, as the crater bottom moves deeper into the sample.

The sputtered atoms move by diffusion into the negative glow region of the plasma. We assume that this diffusion process does not destroy the close relationship between the composition on the sample sub-surface and the number of atoms reaching the negative glow. We expect the diffusion process to be in dynamic equilibrium.

The negative glow is characterised by a relatively high density of charged particles and excited argon atoms. Using different processes, such as inelastic collisions and charge transfer, some of the sputtered sample atoms are excited. The excited atoms then relax to lower energy states, emitting characteristic photons that can be detected by the spectrometer. Again, we assume that the excitation and emission

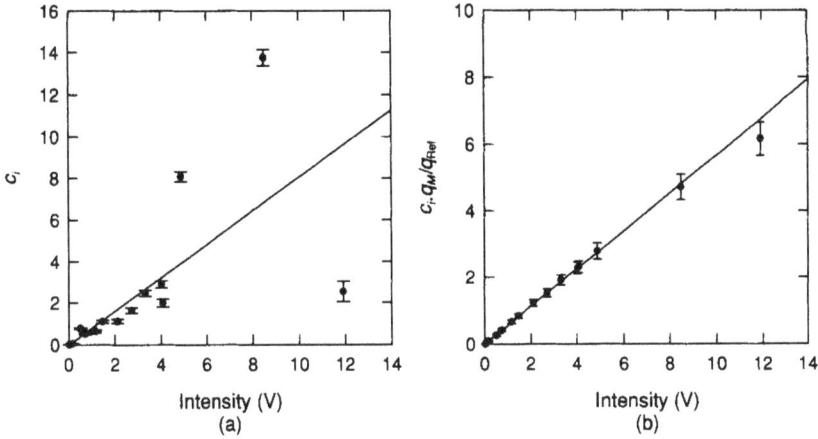

Figure 6.1 *Multi-matrix calibration for Si 288, presented with ordinate: (a) content and (b) content multiplied by relative sputtering rate*

process happens in a way that the number of emitted photons gives a clear picture of the atoms in the plasma, and by deduction of the sample sub-surface composition.

The density of argon atoms in the plasma is much higher than the density of sputtered sample atoms, so we expect the plasma to be characterised mainly by the argon with the presence of sample atoms acting only as a small perturbation. That is, we ignore matrix effects on the nature of the argon plasma, at least, in our initial understanding.

Finally, we assume that the number of photons detected by the spectrometer is representative of the number of photons emitted by the sputtered sample atoms. As the sample atom density is fairly low, we expect that self-absorption will not be too bad.

As a consequence, we expect that the intensities recorded in the spectrometer are proportional to the atom densities in the plasma; these in turn are proportional to the product of the composition of the sample sub-surface and the sputtering rate.

Happily, in practice, most of our assumptions hold fairly well and we often see a nearly linear correlation between detected signals and the product of composition and sputtering rate. An example, showing the effect of sputtering rate on Si 288 in a multi-matrix calibration (described later) is shown in Figure 6.1. Deviations from this simple model can be treated as perturbations.

1 Sputtering Rate

The sputtering rate describes the rate at which material is removed from the sample through particle bombardment. In GDOES we usually use mass sputtering rate expressed in mass per unit time or in mass per square metre per second. The rate at which material is removed from the sample depends on the material and the plasma conditions. This is described in more detail in Chapter 11.

Two different ways of determining the mass sputtering rate are commonly used:

- the mass loss of the sample after a given sputter time
- the volume of the sputter crater after a given sputter time

Instead of the absolute sputtering rate for each sample, the sputtering rate can also be expressed relative to the sputtering rate of a common material, such as pure iron, measured under the same conditions. This has the advantage that such values can be compared with different instruments, even if the anode diameter is not exactly the same or the power, current, voltage, *etc.* are not identical.

Measuring Mass Loss

The mass loss of the sample after sputtering can be measured by directly comparing its mass before and after sputtering. The typical masses of common calibration samples vary from some tens of grams to hundreds of grams. The mass loss using a 4 mm anode is usually less then a milligram per minute. The balance used for this kind of measurement obviously requires very good precision.

Normally, a crater depth of 20–40 μm is chosen. For steel, with a 4 mm anode this corresponds to a total mass change of 2–4 mg. If the desired precision is ±5%, this corresponds to ±10–20 μg. But if five craters with identical sputter times are made on a single sample, the mass change is 10–20 mg, and the precision required is more manageable at ±50–100 μg. Remember that many digital mass balances have a one digit reading uncertainty and a one digit calibration uncertainty, giving a combined uncertainty of ±2 in the final digit. Also, electronic balances are disturbed by magnetic fields, so steel samples should be demagnetised before weighing.

Handle the sample carefully with gloves. Before beginning, check that you can weigh the sample, mount it on the source, unmount it without sputtering, and then reweigh the sample, all without changing its mass significantly. If you cannot do this, then there is little reason in continuing.

To improve the accuracy of the results, the measurements could be repeated for different sputtering times. The sputtering rate could then be calculated by regression. Another possibility to improve accuracy is by using Boumans' equation, or similar equations described in Chapter 11. The mass loss is measured for different excitation conditions, for a fixed sputtering time,[2] and the sputtering rate constant is again calculated by regression.

The uncertainty of the mass loss can be determined using the rules of uncertainty propagation. The mass loss w_l is given by:

$$w_l = w_b - w_a \tag{6.1}$$

$$s(w_l) = \sqrt{s^2(w_b) + s^2(w_a)} \tag{6.2}$$

where w_b and w_a are the sample mass before and after sputtering, respectively, and s is the estimate of the standard deviation.

It is assumed that the amount of material redeposited on the edges of the crater is not significant. Estimates from crater profiles, such as those shown in Figure 6.2

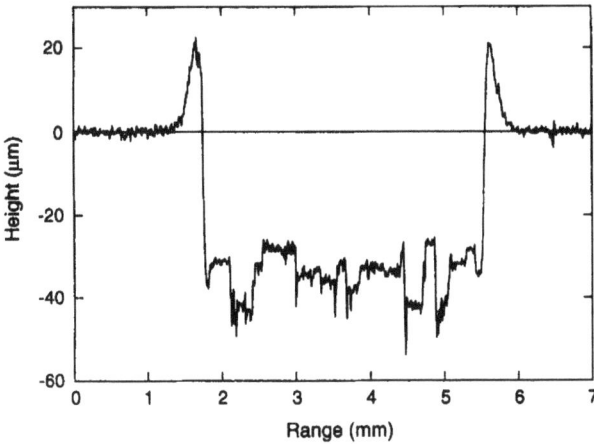

Figure 6.2 *Sputter crater formed on a pure iron sample, showing redeposited material on the sample surface at the edges of the crater, and a flat crater shape with large variations in crater depth due to differential sputtering in the large grains in the sample*

and later in this chapter, suggest it is about 7% of the total mass sputtered from the crater. Redeposition on the crater edge is therefore a possible source of systematic error, though it should be less significant for relative sputtering rates (RSRs).

Determining Crater Volume

The sputtered volume is measured directly from the width and depth of the crater using a profilometer. This procedure also allows checking the flatness of the crater bottom.

For a cylindrical anode, the sputtered volume V is given by:

$$V = Ad = \pi r^2 d \tag{6.3}$$

where A is the crater area, r is the crater radius and d is the average crater depth. When relative sputter rates are used, the crater surface area (πr^2) cancels as it is assumed to be the same for both samples.

From the definition of density ρ:

$$\rho \equiv M/V \tag{6.4}$$

where M is mass, the sputtering rate (mass/s) is:

$$M/s = \rho V/s = \rho Ad/s \tag{6.5}$$

and the sputtering rate per unit area per second is:

$$M/A/s = \rho \, (d/s) \tag{6.6}$$

The depth per unit time (d/s) is called erosion rate. The RSR is simply:

$$(q_M/q_{Ref}) = \frac{\rho_M \, (d/s)_M}{\rho_{Ref} \, (d/s)_{Ref}} \tag{6.7}$$

To reduce the measurement uncertainty, it is sensible to repeat the measurement for several different craters and for different sputter times or plasma conditions, and calculate the average erosion rate through regression.

For most samples and most common profilometers, it is sensible to burn craters to a depth of about 10–30 μm. Smaller depths are usually difficult to measure precisely and larger depths can produce systematic changes in the plasma and hence in the sputtering rate with depth.

Measuring Crater Depth

Several types of instruments are used to measure crater depths. Amongst these are profilometers, also called surface roughness instruments, and interferometers. The most common profilometers use either a contact diamond stylus, optical focus, or laser. Some profilometers are capable of recording the whole crater but many are capable of only a single trace across the crater.

A full crater is shown in Figure 6.3. To determine the average depth of the crater, it was first necessary to level the slope of the sample (in the profilometer software) and then identify the areas inside (dark) and outside (light) the crater. The difference in height for the two regions in Figure 6.3 was 16.8 μm.

The crater depth can also be estimated from 2D scans across the crater. For the crater in Figure 6.3, the 2D scan gave an average depth of 16.9 μm, in good agreement with the 3D scan. Because of the roughness of the crater bottom, it is a good practice with 2D scans to do at least two scans per crater, rotating the sample each time, to scan a different section of the crater.

3D and 2D scans usually give similar results, as it did here, when the crater bottom is nearly flat. But they will not generally be the same if the crater bottom is not flat.

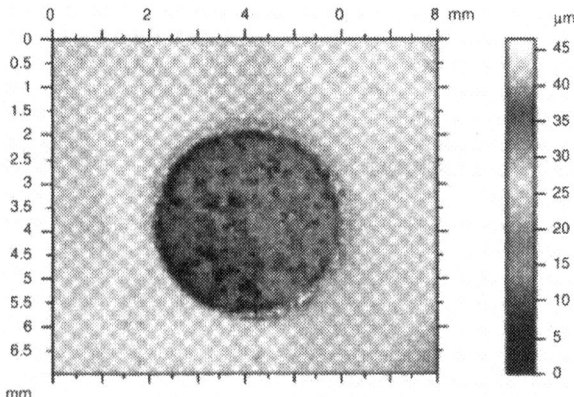

Figure 6.3 *Sputter crater formed on a pure iron sample, recorded with a scanning laser profilometer*

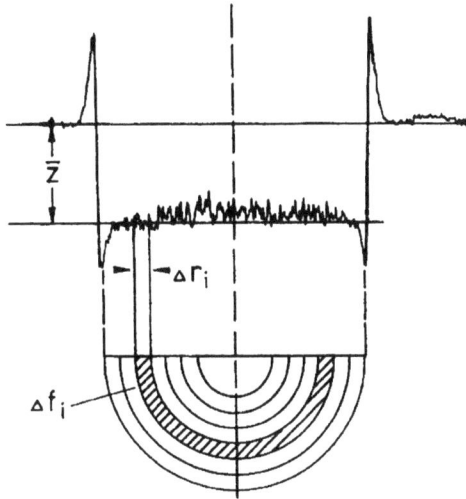

Figure 6.4 *Difference between a line scan and an area scan, where $\bar{Z}$ is the mean depth, Δr_i is a linear segment in the line scan, corresponding to an area Δf_i in an area scan*[3]
(Reproduced with permission from A. Quentmeier, in *Glow Discharge Optical Emission Spectrometry*, R. Payling, D.G. Jones and A. Bengtson (eds), John Wiley & Sons, Chichester, 1997, 295–9)

This is illustrated in Figure 6.4. The 'line scan' shown represents a single scan across the crater; different area segments of the crater are shown by semicircles in the bottom of the figure. Clearly, equal linear segments of the line scan represent very different crater areas. A single scan overemphasises the centre of the crater.

The surface weighted average, which should be used for the sputtering rate calculation, is given by:

$$\langle h \rangle = \frac{1}{2\pi r_0^2} \int_{-r_0}^{r_0} \pi h(r) r \, dr \approx \frac{1}{2S} \sum_i \pi r_i \, \Delta r_i h(r_i) \tag{6.8}$$

Ideally, the scan should be across the centre of the crater. In this case r_0 is the crater radius and S is the crater area. But this is not always achieved in practice, so r_0 then becomes simply half the crater width given by the scan, and is generally less than the true crater radius.

If it is not possible to make this calculation with the available apparatus, it may be adequate to use the depth at the weighted average radius. This average radius is given by:

$$\langle r \rangle \approx \int_{-r_0}^{r_0} \pi r^2 \, dr \bigg/ \int_{-r_0}^{r_0} \pi r \, dr = \frac{2}{3} r_0 \tag{6.9}$$

Note that Equation 6.9 gives the exact value for a 'V' shaped crater bottom.

How the position of the average depth can be estimated in a 2D scan is illustrated in Figure 6.5: (a) for convex craters, the weighted average (solid line) is slightly higher

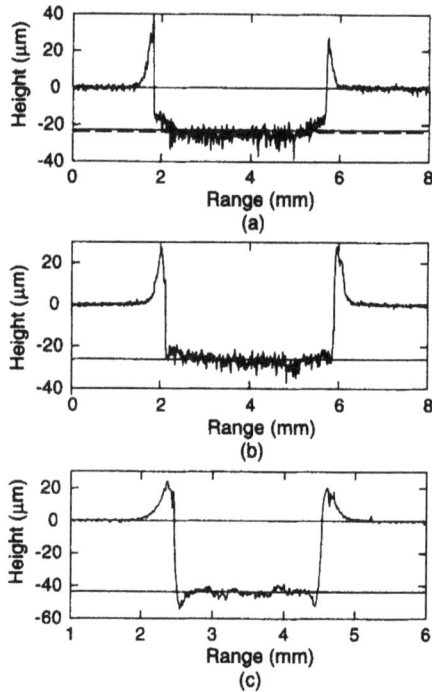

Figure 6.5 *Illustration of where to measure the depth for various crater shapes: (a) convex, (b) nearly flat and (c) concave, where (a) and (b) are for 4 mm anodes and (c) is for 2 mm anode. Solid line: surface weighted average and dashed line: 2D average*

than the non-weighted 2D average, (b) and (c) for nearly flat and concave craters, the weighted and non-weighted estimates are nearly the same. In (a) the solid line is at 22.6 µm and the dotted line is at 23.8 µm, a difference of 5%. Non-weighted 2D scans will therefore tend to slightly overestimate the sputtering rate for convex craters.

Estimating 'Sputter Factors'

It is also possible to calculate the RSR of a calibration sample using the CDP calibration itself and then reinject this sample into the calibration with the new RSR. In effect, the RSR of the sample is altered to ensure the sample fits on the calibration curve(s) for one or more major elements. The sample with its new RSR can then be used for calibrating elements, other than the ones used to estimate the RSR, and in other calibrations. This process can be automated in software.

As an example, consider a new stainless steel sample. We assume the major elements Fe, Cr and Ni have already been calibrated using other samples. We give our new sample a relative sputtering rate of 1. We measure the Fe, Cr and Ni intensities in our new sample and from the calibration curves, we obtain a composition of 120%, when we know from the certificate that it should be 95%. The RSR for this sample

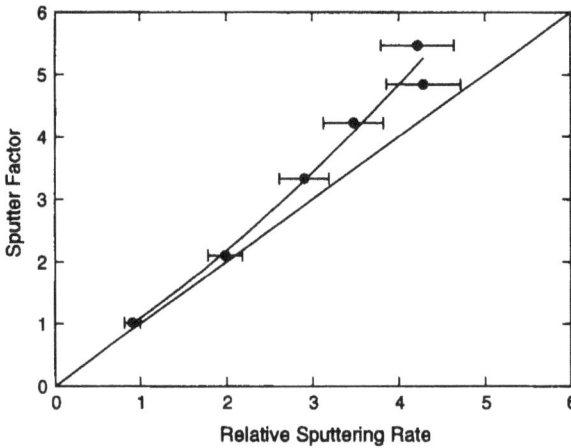

Figure 6.6 *DC sputter factors from ref.[5] compared with measured DC sputtering rates relative to pure iron, for six certified Zn–Al alloys*

must therefore be $120/95 = 1.26$. This sample can now be used to calibrate minor elements, such as Co, Mo or Mn. Relative sputtering rates calculated in this way are called 'sputter factors'.[4]

Using this procedure it is important to estimate the uncertainty in the sputter factor, to keep the possibility of estimating the uncertainty in the final analytical result. Also, when this sample is used to 'improve' the calibration curves for elements already used to calculate its RSR, care must be taken in interpreting the quality of such calibration curves, as most regression procedures assume independent data points, which is no longer the case.

Sputter factors may also show systematic bias for parameters not included in the regression. Figure 6.6 shows the sputter factors (relative to iron) determined for six Zn–Al alloy CRMs, compared with measured relative sputtering rates (relative to iron). The sputter factors are those reported on the certificates for these materials. The RSRs were measured with DC, at constant 1000 V and 25 mA, with varying pressure. Although the uncertainties in the measurements are large, the sputter factors appear systematically too high for increasing RSRs. Unless there is a systematic error in the measured rates, the result is consistent, for example, with a pressure effect on the emission yield, not included in the calibration used to calculate the sputter factors. With constant current and voltage, Zn–Al alloys operate at lower pressures than iron; since emission yields tend to increase slightly with lower pressure, the higher emission yield is then wrongly interpreted as an increase in sputter factor. Since different lines of the same element will have different responses to changing pressure, they will show different sputter factors, though, of course, the RSR of the element is the same.

Measured Relative Sputtering Rates

The RSRs of different materials can vary enormously from one matrix to another. It is good to be aware of typical sputtering rates for different materials as this can

Table 6.1 *Sputtering rates for some nearly pure solid elements*, relative to iron, where range shows the range of n separate measurements*

Element	Z	RSR Mean	Range	n
Al	13	0.37	0.34–0.39	2
Ag	47	9.3		1
Au	79	8.1	5.0–11	2
Co	27	1.8	1.2–2.4	2
Cr	24	1.0	0.77–1.1	3
Cu	29	3.5	3.4–3.6	5
Fe	26	1.0		Ref
Mo	42	1.3	1.2–1.4	2
Nb	41	0.71		1
Ni	28	1.5	1.49–1.52	2
Pb	82	17		1
Si	14	0.21	0.17–0.25	2
Sn	50	6.5		1
Ta	73	3.4		1
Ti	22	0.43	0.427–0.430	2
V	23	0.50		1
W	74	2.9	2.5–3.3	2
Zn	30	8.2	7.8–8.6	2
Zr	40	0.77	0.50–1.0	2

* Collected from unpublished results shared within the GDOES community, principally from (in alphabetical order) M. Aeberhard, C. Autier, T. Asam, K. Crener, M. Köster, T. Nelis, R. Payling and C. Xhoffer.

affect the optimum conditions for analysis and the interpretation of results, especially sputter depth profiles. Table 6.1 shows the measured RSRs for a range of nearly pure elements. They vary from about 0.2 for Si to 17 for Pb; values given here are for nearly pure iron as reference material.

Table 6.2 shows measured RSRs for a range of common materials. Some materials, such as Cu–Al alloys and cast iron, show relatively small variations, while others, notably Cu–Zn and Zn–Al alloys, show large variations depending on composition.

Measured values for RSRs can vary significantly from one laboratory to another and from one operator to another. Table 6.1 highlights the range of values that can be obtained by different laboratories, even for simple, single element solids. Although the measured values vary, it is thought that the 'real' values do not vary to such an extent. Different laboratories use different methods and different apparatus (microbalances, 2D and 3D profilometers with stylus or laser, *etc.*) and there is as yet no internationally agreed standard for measuring and reporting RSRs. We have also found no indication of the associated uncertainties in values circulated between laboratories.

Much of the variation is thought to be due to different ways of estimating crater depth, to compensate for changing crater shape, *etc.* It may also be due to insufficient number of craters being used for each measurement. If an operator measures only a single crater, and there is an unnoticed incident during sputtering, perhaps caused by dust on the sample surface leading to a temporary short circuit or an inadequately

Table 6.2 *Relative sputtering rates (RSR) for a large number of CRMs*, arranged into material types (families), where n is the number of CRMs measured in the family*

Major element	Material type	Typical CRM	Typical composition (mass %)	Typical range of RSR	n
Al	Al/Mg	Pechiney 6039	Al 95 Mg 4 Fe 0.6	0.55–0.58	3
Al	Al/Si	MBH G28J5	Al 71 Si 26 Mn 0.9	0.39–0.56	17
Al	Al/Zn	Pechiney 9165	Al 90 Zn 6 Mg 2 Cu 2	0.66–1.3	2
Al	Ceramic	SIMR CC650A	Al 37 O 32 Ti 22 C 5	0.19	1
Co	Alloy	MBH X404C	Co 53 Cr 24 Ni 11 W 7	2.4	3
Cu	Brass	CTIF LH11	Cu 67 Zn 26 Ni 3 Pb 1	2.2–4.6	32
Cu	Bronze	CTIF UE51/26	Cu 82 Sn 7 Zn 4 Pb 4	4.9–5.1	2
Cu	Cu/Al	CTIF 2151/R	Cu 85 Al 9 Fe 4	2.1–2.3	4
Cu	Cu/Ni	MBH C62.13	Cu 84 Ni 14 Mn 1	2.3	1
Cu	Cu/Si/Zn/Fe	MBH WSB-4	Cu 87 Zn 5 Si 4 Mn 2	2.7	1
Cu	Low alloy	MBH 17868	Cu 99 P 0.1 Cr 0.1	3.5	5
Fe	Cast	CTIF FO2-2	Fe 93 Si 3 C 2 Mn 1	0.71–0.97	41
Fe	Coronite	SIMR JK41-1N	Fe 48 Ti 25 Co 7 N 7	0.92	1
Fe	High-alloy steel	MBH BS 186	Fe 63 Ni 36 Mn 0.7	0.92–1.8	4
Fe	Low-alloy steel	NBS 1763	Fe 95 Mn 2 Si 0.5	0.90–1.1	60
Fe	Stainless steel	BAS 464	Fe 52 Cr 26 Ni 21	1.0–1.4	23
Fe	Tool steel	BAS 486	Fe 81 W 6 Mo 5 Cr 5	1.0–1.7	9
Mg	Mg/Al/Zn	MBH MGB1B	Mg 95 Al 3 Zn 2	1.1	1
Ni	High alloy	MBH 4005B	Ni 70 Cu 21 Si 2 Al 2	1.3–1.9	11
Ni	Low alloy	BRAMMER BS 200-1	Ni 99.6 Mn 0.1	1.5–1.6	6
Ni	Ni/Cr/Co/Mo	MBH 14939 D	Ni 48 Cr 21 Co 20 Mo 6	1.3–1.5	9
Sn	Low alloy	MBH SR3/A	Sn 99 Pb 0.3 Sb 0.2	6.5–6.6	3
Sn	Solder	MBH S63PR2	(Sn 50 Pb 50)	5.5	1
Ti	Various	NBS BST-15	Ti 88 Al 6 Mo 3 Cr 1	0.46–0.62	6
Zn	Low alloy	MBH 41Z3G	Zn 99.95	4.5–7.8	9
Zn	Zn/Al	MBH 42ZN5E	Zn 96 Al 4 Cu 0.1	1.2–5.9	23

* Collected from unpublished results shared within the GDOES community, principally from (in alphabetical order) M. Aeberhard, T. Asam, K. Crener, T. Nelis, R. Payling, R. Toutain, Z. Weiss, K. Wilsdorf and C. Xhoffer.

cleaned anode changing the effective anode-to-sample distance, then a poor value for the relative sputtering will result, and measuring a second or third crater would have shown the problem. Measurement of several craters also allows some estimation of the uncertainties of the measurements.

2 Density

The density of the sample is needed when the erosion rate is used to measure the sputtering rate. The density can be measured directly by the Archimedes method, taken from tables when the sample is a well-known alloy or compound, from sample geometry and mass, or calculated using approximations. The uncertainty in the density value should be smaller than other uncertainties associated with the calibration, so $\leq 1\%$ is desirable and $\leq 3\%$ is acceptable.

Use of tabulated values is certainly sensible as it saves considerable time, and the authors of such tables have presumably put much effort into measuring their results. However, some care should be taken as the density of some materials can vary significantly depending on the manufacturing process. An example is cast iron, where C can be present in different forms depending on its heat treatment.

When the sample has a regular shape, *e.g.* a cylinder, the volume can be estimated from the sample dimensions. The uncertainties can then be estimated by the principle of uncertainty propagation.

Probably, the fastest and easiest way to measure density is with a pycnometer, a container that can be filled with an exact amount of water. The operator first weighs the pycnometer filled to a predefined level with deionised water at a measured temperature, and then partly empties the pycnometer. Then the operator weighs the sample and places it in the pycnometer. They then fill the pycnometer with water back to the predefined level and reweigh the pycnometer. The change in weight of the pycnometer minus the mass of the sample is the mass of the displaced water. When this value is divided by the density of water at the measured temperature (the density of water at 20 °C is 998.231 kg m^{-3}, and decreases by about -0.208 kg m^{-3} per degree near 20 °C), this gives the volume of the sample. The mass of the sample divided by the volume is the density of the sample.

Measurement of density using the Archimedes method is fairly safe. At least, one can control the measurement and estimate the uncertainties. Density kits are available for many mass balances, and details on the Archimedes method can be found in many physics books. The operator first places a beaker partially filled with deionised water at a measured temperature on the balance, and tares the balance. Then it suspends the sample in the beaker so that it is fully submerged in the water. Care is needed to ensure that air is not trapped around the sample, and that the sample is stationary and not touch does the beaker. The difference in weight between the beaker plus water plus sample, and the beaker plus water is the mass of the displaced water. This value divided by the density of the water at the measured temperature gives the volume of the sample. The mass of the sample divided by the volume is the density of the sample.

When the density of metallic alloys is to be determined, the 'hard-core' model gives a fairly good approximation. This is described in Chapter 11. The hard-core model is particularly successful when alloys are made from elements of roughly the same size, such as Fe, Ni and Cr, or when atomic bonds are not too strong. Significant amounts of atoms with very different sizes, *e.g.* Pb, W, Al, C, *etc.*, will deteriorate the accuracy of the approximation.

Uncertainty of Sputtering Rate Corrected Composition

It is now generally acceptable, in CDP calibration, to replace the composition c_i by the product $c_i q$, where q, depending on the software, is either the absolute sputtering rate or the RSR of the calibration sample. This is the so-called cq approach. Use of absolute or RSR does not change anything in terms of treating the uncertainties of the measurement. The uncertainty is estimated by:

$$\sigma(c_i q) = \sqrt{q^2 \sigma^2(c_i) + c_i^2 \sigma^2(q)} \qquad (6.10)$$

It should be mentioned that the uncertainty in the sputtering rate of a CRM may be significantly larger than the uncertainty in the certified composition. This is particularly true for the major elements. At best, the relative uncertainty of a measured sputtering rate is about 5%; often it is 10% or worse. Measured sputtering rates then become the largest source of uncertainty in CDP calibration.

3 Linearity

Some older calibration methods for CDP require calibration curves to be linear. In particular, linearity is required when sputtering rate corrections are applied to intensities, rather than to composition. In the '*cq*' approach, calibration curves do not need to be linear. A few remarks, however, need to be made about linear and non-linear calibration curves.

The non-linearity observed for some calibration curves in GDOES is generally caused by self-absorption, see, for example, Al 396.401 and Cr 425.433 in the multi-matrix calibration in Figure 6.7. This non-linearity affects resonance lines or near-resonance lines, *i.e.* emission lines with a zero or low final excitation level. These lines are generally very sensitive and therefore good for analysing low elemental compositions. For higher compositions their sensitivity decreases significantly, *i.e.* the slope or local derivative increases. This implies that large changes in composition, at the high end, will produce only small variations in intensity. In other words, the uncertainties of the composition ($c_i q$) will increase faster than the uncertainties of the intensities.

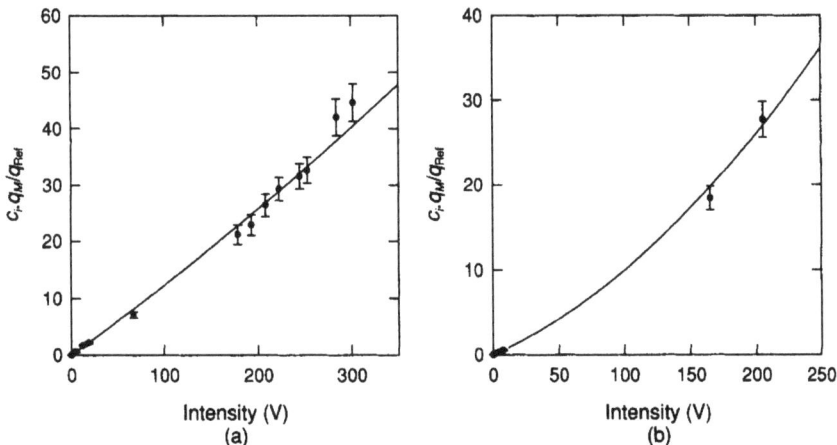

Figure 6.7 *Multi-matrix analytical curves for: (a) Al 396.401 and (b) Cr 425.433*

On the other hand, lines that are linear over a wide composition range, and give excellent results for higher compositions, often lack sensitivity for low elemental compositions and traces.

Though it is possible to cover the entire composition range, from 0% to 100%, with only one spectral line, better analytical results, in terms of precision and accuracy, are obtained when adequate lines are used for each composition range. It is certainly a sensible practice to include several lines of the same element during calibration and choose the best one during analysis.

Extrapolation, the use of a calibration curve beyond its calibrated range, should be avoided. However, if extrapolation is necessary because no adequate calibration samples are available, it should be done only with linear calibration curves.

4   Matrix Effects

By definition, matrix effects are a 'change in the intensities or spectral information per atom of the analyte arising from a change in the chemical or physical environment',[6] *i.e.* they are changes in elemental intensity not simply related to changes in composition of the element. They could include changes in intensity due to changes in sample shape or size, changes due to sample morphology, sample temperature or conductivity, changes related to other layers in the sample and changes caused by other elements, including changes to the plasma by sputtered atoms.

For convenience, in GDOES, matrix effects are separated into two groups:

- additive effects, usually treated as spectral interferences
- non-additive effects, due to changes in emission yield

Changes in sputtering rate, which are dealt with by the '*cq*' approach, will not be considered here. The difficulty in matrix effects on varying emission yields is that the emission yield is a characteristic of each spectral line and not just of the element and the plasma.

Effects of Changes in Plasma Impedance

Varying plasma impedance affects emission yields. A well-known example is the effect on the emission yield for the Si 288 line. In constant power/constant pressure mode, the plasma impedance varies depending on whether Si is analysed, for example, in an Fe-based or a Zn–Al-based material. The effect is well described in the literature and is generally understood to be a consequence of significant differences in the secondary electron emission yields of these materials (see Chapter 11 for more details). The change in plasma impedance causes changes in voltage and current, which then influence the emission yield of the Si line. Other lines, from other elements, show this effect to greater or lesser extent depending on the combination of materials used in the calibration.

Different ways exist to correct this problem. The first is to maintain the plasma impedance constant by varying the plasma pressure. This leads to the constant voltage/constant current approach.[7] This may introduce a matrix effect on emission yields caused by varying pressure, but generally this is considered to be small.

A different solution lies in monitoring the changes in impedance, *e.g.* by monitoring the variations in voltage or current, establishing the effect of these variations on the emission yield and then correcting it numerically.[8]

Which approach is 'better' is still being discussed in the literature, but each certainly has its advantages and disadvantages.

Other Matrix Effects

Matrix effects have other ways of influencing the emission yield. Many different reactions occur simultaneously in the plasma. The probability of these reactions is dependent on the composition of the reacting species and the reaction cross-section. If the reaction cross-section between two species is particularly high, the content of these species in the plasma may be significantly decreased by the reaction. This seems to be the case when hydrogen is added to the plasma. The details are not yet completely understood, but hydrogen can strongly influence the emission yields of other atoms. A more detailed discussion is given in Chapter 11. As most of the atoms in the plasma are argon atoms, we would expect such 'chemical effects' on the plasma to be small and treatable by small correction terms. Hydrogen, with its very strong interactions with an argon plasma, is an exception. Nevertheless, as we go into more and more detail in our understanding of analytical glow discharges, we have to expect to encounter more of these effects in the future.

5 Optimisation of Calibration Curves for CDP

While the remarks on calibration in Chapter 4 are fully applicable to CDP calibration, there are a few additional points to consider related specifically to CDP, or multi-matrix, calibration.

Choice of Reference Materials

It is still a relatively common practice to use only a few reference samples for calibrating a CDP method. Although in some cases it may be possible, and justifiable in saving time or cost, it usually leads to significant uncertainties in the final results. We, therefore, clearly recommend the use of many CRMs to establish CDP calibration curves, only then is it possible to check the linearity of the calibration functions, check variations in spectral background, small matrix effects, *etc.* In addition, the use of multiple CRMs addresses the uncertainties of the sputtering rate measurements.

The effect of larger uncertainties in sputtering rates means that more calibration samples would be needed in CDP calibration to reduce uncertainties in regression parameters to values comparable to bulk calibration of a single matrix. The use of fewer calibration samples in CDP calibration, by some analysts, suggests that they are willing to accept lower accuracy in CDP or are more willing to modify calibration curves by 'mini-calibration', *i.e.* to adjust calibration curves to give good results for known coated samples.

As for bulk analysis, the CRMs should be selected to cover the entire composition range required in the analysis. When choosing the calibration samples for CDP calibration, one should be aware that the important parameter to consider is not simply the composition but composition times the sputtering rate, *i.e.* '*cq*'. For example, for Zn–Ni coatings on steel, with typically 12% Ni in Zn, the calibration should include Ni samples with more than 40% Ni content, to compensate for the very high sputtering rate of Zn–Ni coatings. As a rule of thumb, at least some calibration samples should have higher intensities than the analyte samples. Also, include samples with low intensities for each element, to avoid severe covariance between slope and intercept in the calibration curves.

Background Correction

The spectral background and its origins will be discussed in Chapter 11. In this chapter we will discuss the importance of background effects in CDP analysis. In fact, there are two common mistakes to avoid: the first is linked to non-weighted regression, and the second is linked to the dependence of the spectral background on the material.

Non-weighted Regression

As discussed in Chapter 4, non-weighted regression will minimise the squared deviations of the measured values from the calibration curve. As a consequence, the deviations at high and low contents will be similar. When this is combined with an insufficient number of calibration samples, the deviations at low compositions can become unacceptably high. In Figure 4.7, we showed how log–log plots could identify this problem. Here we provide another example, a CDP calibration for Cr 425, but this time with linear scales. The weighted curve was shown in Figure 6.7b. Unweighted in Figure 6.8a, the calibration curve looks just as good especially at high contents. But a zoom near the origin in Figure 6.8b shows the problem.

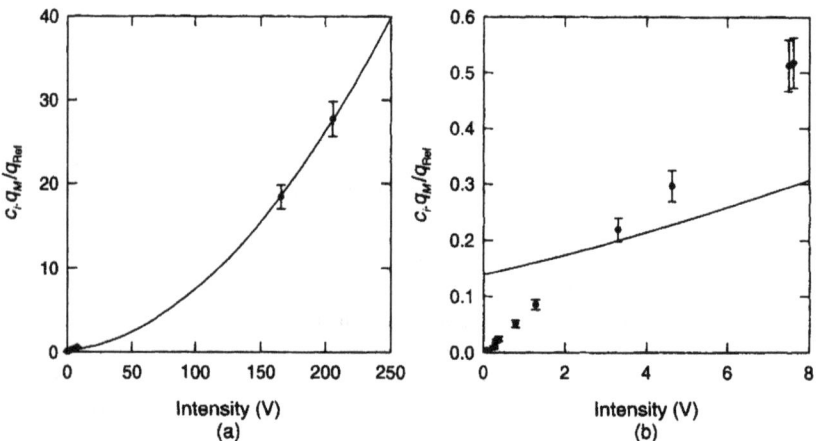

Figure 6.8 *Analytical curves for Cr 425.433 (similar to Figure 6.7 but now unweighted): (a) full range and (b) expanded near the origin*

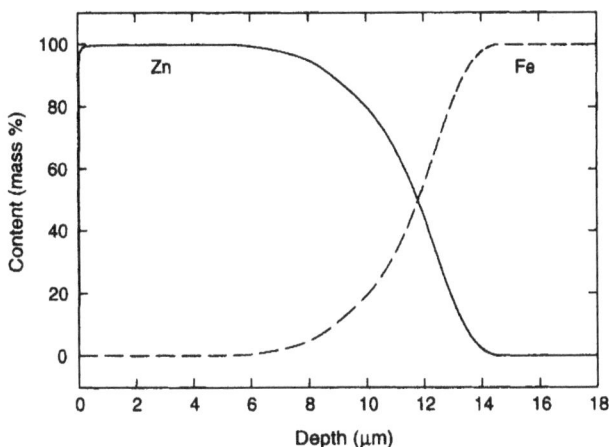

Figure 6.9 *CDP of galvanised steel, where a problem in the Zn 335 calibration causes the Zn to disappear suddenly at about 14 μm. Using equally weighted data, the intercept was too negative, i.e. the BEC for Zn 335 was too large*

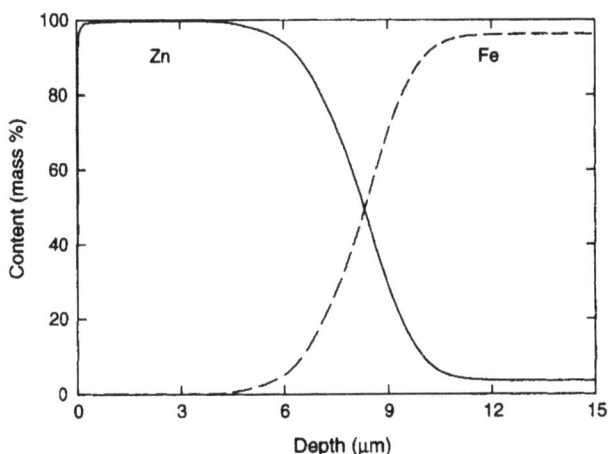

Figure 6.10 *CDP of galvanised steel, where a problem in the Zn 335 calibration causes the Zn to remain too high in the steel. Using equally weighted data and an insufficient number of low points, the intercept was positive, when it should be a small negative number*

Calibration curves with poor estimates at low contents lead to 'tell-tale' problems in compositional depth profiles. Major elements in a coating will either suddenly disappear at an interface or show an unreasonably high content in the base material. These two problems are illustrated in Figures 6.9 and 6.10, respectively.

One solution would be to make a proper estimation of the uncertainties. This would probably show that the effect is not significant, because the uncertainties are too large, due to unsound calibration. A better solution to this artefact is weighted regression and increased number of samples in the calibration curves.

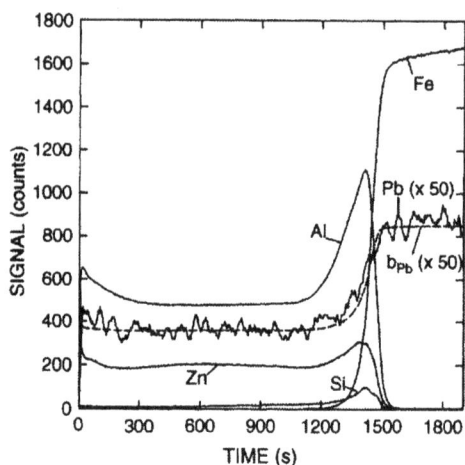

Figure 6.11 *Apparent variation in Pb content due to varying background signals in the coating and substrate*
(Reproduced with permission from R. Payling, in *Glow Discharge Optical Emission Spectrometry,* R. Payling, D.G. Jones and A. Bengtson (eds), John Wiley & Sons, Chichester, 1997, 392–402).

Changes in Spectral Background

The second common 'calibration mistake' is linked to the change of spectral background with the major elements in the sample. The problem typically is noticed when some elements are present in minor compositions in the coating but are not expected in the substrate, or vice versa. Because the element is not expected in one of the materials, few calibration samples would be available or included in the calibration with certified values for this element in this material. Transition metals with open d-shells, such as Fe, Ni and W, show a relatively high spectral background. Zn, on the other hand, has a low spectral background. If we examine a Zn-based coating on steel, the spectral background will change in going from the coating to substrate. This is shown in Figure 6.11 in the Pb profile, where there is no detectable Pb for this sample in either coating or substrate.

The solution to this problem is to introduce an additive 'interference' term to the calibration equation, allowing the calibration function to adjust for the effect. Care must be taken to introduce a sufficient number of reference materials containing a known amount of our element in the different matrices. Sometimes it is convenient to include a pure material of another element with '0 ± uncertainty' of our element of interest.

Check Accuracy with Known Samples

As for bulk analysis, it is necessary to check the accuracy of the CDP calibration by analysing some well-characterised samples. It is sensible to analyse both homogeneous reference materials and coated materials.

The chemical compositions of homogeneous reference samples are well known; and they can often contain a large number of different elements. A CDP analysis will first provide a check on the accuracy of the determined chemical composition. Then the crater depth can be measured and compared with the calculated depth in the depth profile. If the crater depth is within the expected uncertainty (typically, 5–10% of depth), then the calibration is probably working well for the major components of the analysed sample, as uncertainties in composition in CPD tend to accumulate in the depth estimate.

The second step is to check the accuracy with coated materials of known chemical composition and coating weight. Frequently, these materials can be supplied by the same person supplying the 'unknown' analyte materials. They, therefore, represent the accuracy attainable with the unknown samples.

6 Case Study: Galvanised Steel

The aim is to analyse a range of Zn- and Al-based coatings on steel. ISO has prepared a standard for the analysis of such coatings by GDOES.[10] The first step is to decide the elements to include in the calibration and their ranges in composition, so that the calibration covers the full range of possible coatings to be analysed with the method.

The main elements of interest are Zn, Al, Ni, Si and Fe, and some minor elements such as Pb or Sb. Also, if the surface of the coating is of interest, additional major elements are H, O, C, P and Cr. Since we are interested in going through the coating and into the steel substrate, the range of interest for some elements, particularly Zn, Al and Fe, is 0–100%; and even for elements such as O, C, P and Cr, *etc.*, it may be as high as 40% at the surface. So, we need to choose our calibration samples, where possible, to cover all these ranges.

If the intention is to analyse a wide range of Zn-based coatings, then it is good to include a large variety of reference material types in the calibration, representing all the possible alloys present in the coatings and steel substrate. Normally, calibration samples are chosen from low alloy steel, stainless steel, Zn–Al alloys, Al–Si alloys, brass and nickel alloys. A low alloy steel sample is useful in checking the low points in the calibrations of the elements, except, of course, for Fe. It is also useful as the sputtering reference.

Though the maximum Ni content in Zn–Ni coatings is about 12%, the samples with high nickel contents are included because Zn–Ni coatings have an extremely high sputtering rate so that the Ni content of 12% gives signals comparable to high nickel alloys. No certified Zn–Ni materials were available. Though it was not done here, it is also interesting to include some well-characterised coated steel samples. They can be included in the calibration when they have accepted values for their coating composition and thickness or coating weight. They are also a useful '*in situ*' tool for estimating the accuracy of the calibration curves.

Although high Cu compositions are not of major interest in these coatings, the brass and copper alloy samples were included because they can be used in setting up the conditions for good crater shape; they also provide intermediate Zn intensities, and help with matrix corrections such as DC bias voltage correction. Brass samples are also useful for estimating the spectral background.

Table 6.3 *Calibration samples used for calibration of Zn–Al coatings*

Sample	Type	Content	Relative sputtering rate
1045	Al–Si	Al 90 Si 8	0.32
L3	Brass	Cu62 Zn33 Sn2	3.15
WSB6	Cu alloy	Cu94 Si2 Zn1 Pb1	2.63
CKD 242	Cast iron	Fe93 Si3 C2	0.90
CKD 243	Cast iron	Fe93 Si2 C2	0.90
FO5/3B	Cast iron	Fe93 C2 P1 Si1	0.74
BS 4C	Cast iron	Fe95 C4	0.73
JK41-1N	High nitrogen	Fe48 Ti25 N7	0.92
CC650A	High oxygen	Al37 O32 Ti22	0.19
S30APR3B	Lead	Pb66 Sn32 Sb1	5.42
1763	Steel-low alloy	Fe95 Mn2	1.04
1765	Steel-low alloy	Fe99.6	1.00
1766	Steel-low alloy	Fe99.8	1.00 Ref
MW27	Steel-high alloy	Fe45 Ni35 Cr16	1.18
MW37	Steel-high alloy	Fe59 Cr25 Ni12	1.05
43Z11A	Zn–Al	Zn89 Al 11	2.91
43Z13A	Zn–Al	Zn90 Al8	2.37
43Z15A	Zn–Al	Zn91 Al7	2.44
43Z21A	Zn–Al	Zn74 Al24	1.58
43Z23A	Zn–Al	Zn67 Al30	1.24

For our calibration we chose the samples listed in Table 6.3. Unfortunately, no higher Ni sample was available. If the interest was just in the Zn coating, then some samples could have been omitted. The cast iron samples and the high N and high O samples, for example, were included because of our additional interest in the extreme surface of the coatings.

The second task is to choose the emission lines, or if they are fixed on a poly-chromator, understand their nature by referring to Appendix A. For Zn, the most commonly used lines in GDOES are at 213.85, 334.50 and 481.05 nm. The line at 213.85 nm is a strong resonance line and shows severe self-absorption at high Zn compositions, so it is not recommended for analysing Zn-based coatings. The other two lines are not resonance lines and normally give linear calibration curves up to 100% Zn. The only commonly used Al line is at 396.15 nm. This is a strong near-resonance line and subject to moderate self-absorption, but the effect is off-set to some extent by the lower relative sputtering rates of Al-containing alloys and by the moderate source power selected below, reducing the absolute sputtering rates.

The lines we have chosen are shown in Table 6.4. The Fe 371 line, though a resonance line, is only weakly self-absorbing, and normally assumed to be linear. The Si 288 line is a non-resonance line, while the Ni 341 line is a near-resonance line and non-linear. No significant spectral interference is expected with any of the lines, as many possible interferants, such as Co on the Ni 341 line, are not present in the samples.

Table 6.4 *Emission lines selected for major elements in Zn–Al coated steel*

Line	State	s_L $(\times 10^{-12}\ m^2)$	Linear	Possible interferences
Al 396.152	I r	0.67	No	
C 156.140	I r	0.09	Yes	Fe
Cr 425.433	I R	0.94	No	
Cu 224.700	II	0.91	Yes	
Fe 371.994	I R	0.32	Yes	
H 121.567	I R	0.06	Yes	
N 149.255	I	0.11	Yes	Cu, Cr, Fe
Ni 341.476	I r	0.92	No	
O 130.217	I R	0.08	Yes	
P 178.283	I R	0.24	Yes	
Si 288.158	I	0.62	Yes	
Zn 334.502	I	3.07	Yes	

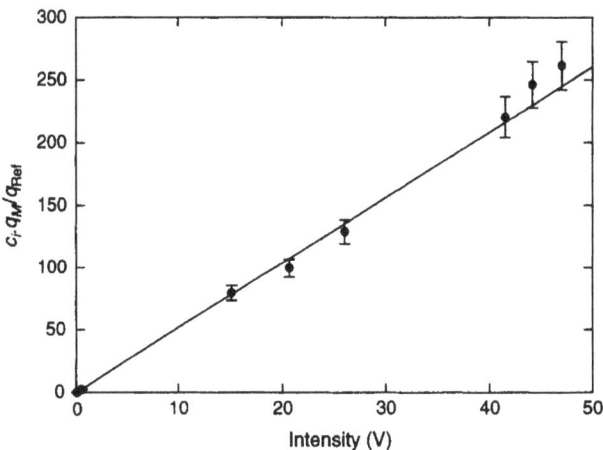

Figure 6.12 *Multi-matrix analytical curve for Zn 334*

In selecting the source conditions, it is important to consider that some Zn alloys have relatively low melting points; it is therefore usual to select moderate conditions. For RF operation this means an applied power of about 35 W, and for DC operation, a current of about 20 mA. The actual values in RF or DC will vary from instrument to instrument. Much of the heating of the sample comes from sputtering. The conditions were therefore chosen to give a sputtering rate in steel of about 3 μm/min^{-1}, as it is known from experience that this will then give good results for Zn-based coatings. The second source parameter is chosen to give good depth resolution; in RF this may be pressure or voltage, and usually voltage (typically 600–700 V) in DC. One way to optimise the depth resolution for Zn coatings is to optimise the crater shape for brass; another is to optimise the sharpness of the interface on some typical coatings.

An RF applied power of 35 W with an argon pressure of 600 Pa was chosen for our calibration.

Having determined the calibration samples, the emission lines and the plasma conditions, the calibration can then be carried out. The resulting calibration curve for Zn is shown in Figure 6.12. The curves for Si, and Al and Cr were shown previously in Figures 6.1 and 6.7, respectively, and Ni in Chapters 4 and 11.

7 Sequence for Calibration for CDP

The following steps are recommended for calibration for compositional depth profiling:

Create method:
1. select elements
2. select calibration, recalibration and sputter reference samples
3. check that composition ranges of calibration samples match analysis needs
4. select source conditions, ensuring good crater shape

Calibrate:
1. measure or calculate densities of samples in method
2. measure sputtering rate of sputter reference sample
3. measure or calculate RSRs of the calibration and recalibration samples
4. make calibration
5. optimise regression
6. check drift-correction (recalibration) standards
7. validate calibration

References

1. B.V. King and R. Payling, in *Glow Discharge Optical Emission Spectrometry*, R. Payling, D.G. Jones and A. Bengtson (eds), John Wiley & Sons, Chichester, 1997, 273–87.
2. R. Payling, *Surf. Interface Anal.*, 1994, **21**, 791.
3. A. Quentmeier, in *Glow Discharge Optical Emission Spectrometry*, R. Payling, D.G. Jones and A. Bengtson (eds), John Wiley & Sons, Chichester, 1997, 295–9.
4. Z. Weiss, *J. Anal. Atom. Spectrom.*, 1995, **10**, 891.
5. Z. Weiss, *J. Anal. Atom. Spectrom.*, 2000, **15**, 1485.
6. ISO 18115, Surface chemical analysis—vocabulary, 2001.
7. K. Marshall, *J. Anal. Atom. Spectrom.*, 1999, **14**, 923.
8. R. Payling, *Surf. Interface Anal.*, 2002, **33**, 472.
9. R. Payling, in *Glow Discharge Optical Emission Spectrometry*, R. Payling, D.G. Jones and A. Bengtson (eds), John Wiley & Sons, Chichester, 1997, 392–402.
10. ISO 14707, Glow discharge optical emission spectrometry (GD-OES)—introduction for Use, 2001.

CHAPTER 7
Drift Correction

1   What is Drift Correction

Drift correction, also known as recalibration or standardisation, is commonly used in OES to correct for 'minor' changes or drift in the spectrometer. This process is absolutely essential if the instrument is to provide acceptable accuracy over extended periods of time stretching into days, weeks, months and years from the time when the instrument was originally calibrated. Although the term "recalibration" is popular in the industry and is synonymous with drift correction, this term can be somewhat misleading. The instrument is not actually calibrated again, but rather drift correction factors are created and applied to the spectrometer during the drift correction procedure. The original intensity values used to establish the calibration curves remain unchanged. Drift correction factors are created for each and every element to be analysed by the instrument. The factors are created by comparing the *present* emission intensity for a particular element to the *original* intensity for the same element. To perform this comparison, a set of homogeneous (recalibration) samples is measured during the initial calibration process and the same set is measured again during any subsequent drift correction. Often the recalibration (drift) samples are added to the calibration curves, though they need not necessarily be included in the regression.

In most cases the drift in the spectrometer is assumed to be linear with respect to the measured intensities (see Figure 7.1) and can therefore be expressed by a set of two parameters. Using these drift correction factors, we can then estimate mathematically from any *currently measured* intensities what these intensities *would have been* if the instrument had not drifted. In other words, we can estimate what we would have measured as intensities if these samples had been analysed during the original calibration process. We, in effect, always try to go back to 'day one' when the instrument was calibrated with the certified reference materials and the calibration curves were established.

The corrected intensities $I_{i\text{corr}}$ for the emission line of element i are given by:

$$I_{i\text{corr}} = a_{i0} + a_{i1}I_i \qquad (7.1)$$

where a_{i0} and a_{i1} are the intercept and slope, respectively, of the linear drift function. In the example shown in Figure 7.1, the detected light was 15% higher during the calibration than during the drift correction procedure.

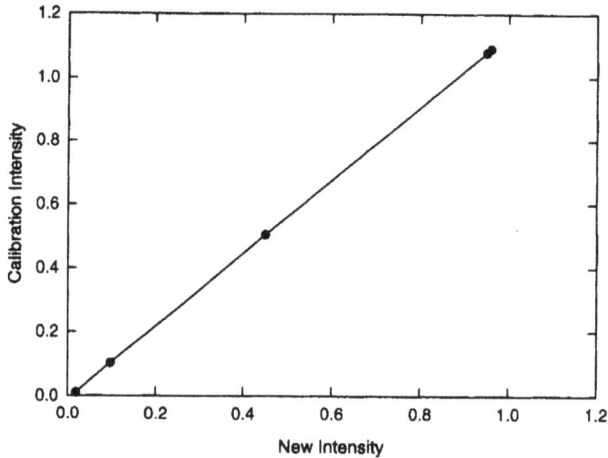

Figure 7.1 *Illustration of drift correction*

In GDOES, instruments may drift, like all emission spectrometers, for various reasons:

• small changes in the source
• small changes in the optical spectrometer(s)

Changes in the source could include slow changes in the anode-to-sample gap caused by wear of the front face of the anode. Changes in an optical spectrometer could include a dirty window, small misalignment of the optics due to temperature or pressure variations, slow ageing of the detector response characteristics and small variations in the gain of electronics.

When large spectral interference corrections are used in the calibration curves, drift correction may lead to significant systematic errors if the line used to measure the magnitude of the interference does not drift in the same way as the analyte line.

Drift correction can be used to adapt the calibration curves to changed instrument settings. For example, if the detector sensitivity is changed, perhaps by changing the voltage on a PMT or the collection time in a solid-state detector, drift correction can work rather well, except of course if the detector is in saturation.

Changes due to altered plasma conditions may also be corrected successfully by drift correction when the calibration curves are linear and the effect of the change on the intensities is linear. An example would be a change in applied RF power for linear calibration lines. The drift correction procedure does not work well, however, when correcting for non-linear changes in plasma conditions, such as pressure or voltage changes, or when non-linear spectral lines are used.

2 Selection of Drift Correction Samples

Drift correction samples should be selected in a way similar to selecting normal calibration samples (see Figure 7.2). The precise chemical composition of the drift

Figure 7.2 *Some materials being considered for drift correction*

sample is not important (unless it is used in the regression) as only its intensities are used to estimate the drift. This does not mean, however, that the chemical composition is of no importance. Drift samples must be very homogeneous, *i.e.* the chemical composition must be the same throughout the entire sample. They should also be thick, if possible, to ensure long life following repeated polishing or machining to remove craters.

To reduce the effect of measurement uncertainties, drift samples should be selected to represent the extremes of the intensity range used for the calibration, *i.e.* at least one acts as a 'high' content sample and one as a 'low' content sample for each elemental emission line. Note that samples must be selected to cover the intensity range, not the composition range, as only intensities are used in the correction, and intensities vary with both composition and sputtering rate.

The selected samples need not be the highest or lowest for each element, otherwise, in many cases, the number of drift samples would be more than the number of elements and nearly as much work would be required as redoing the whole calibration. To minimise the number of drift samples and hence the time it takes to do the recalibration, it is better to choose samples that have relatively high or low contents for many elements. For example, a nearly pure iron sample can be the 'low' content sample for many elements in steel. Ideally, the number of drift samples should be less than five. If one have only a small number of recalibration samples, then one will probably perform a drift correction more often, and this could be good practice if the instrument changes or drifts slowly over time.

3 One Point, Two Point, Multiple Point

As mentioned previously, drift correction consists of a mathematical linear correction to the presently obtained intensities with the purpose of correcting for any minor drift in the instrument. For each elemental emission line, only two samples are needed to establish the correction for background and sensitivity. The process is illustrated in

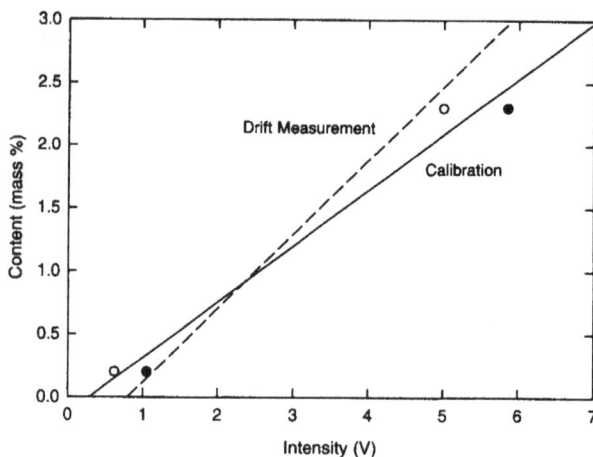

Figure 7.3 *Illustration of two-point drift correction (recalibration):* • *are original calibration values, and* ○ *are recalibration values. Here the background has increased and the intensity has decreased*

Figure 7.3. The drift correction factors are given by:

$$a_0 = \frac{I_1 I_h' - I_h I_l'}{I_h' - I_l'} \quad \text{and} \quad a_1 = \frac{I_h - I_1}{I_h' - I_l'} \tag{7.2}$$

where the primed intensities I' are the new intensities, the unprimed intensities I are the original intensities obtained during the initial calibration, and l and h refer to low and high content samples, respectively. This equation can be written in other equivalent forms, but this is more interesting for the uncertainty calculations that follow.

Because the drift correction process is linear, it is possible to perform a drift correction with only one sample per spectral line, when only one of the parameters of the linear correction needs to be determined. For example, if you are sure that only the slope has changed slightly and not the background signal, measuring a high-content 'recal' sample will be sufficient. Alternatively, if only the background has changed, measuring a low content 'recal' sample is sufficient.

In a multiple-point drift correction, more than two samples are used in the recalibration procedure for each element. This tends to reduce any errors that may be incurred by using just a 'pure' sample for the low point, and a 'high' content sample for the high point in a two-point drift correction. All of the elements in the samples are measured anyway, so it makes sense to use all the measured intensities, and not just those listed as high and low. This would, in theory, provide a more accurate drift correction as more samples are compared for each element.

When more than two drift standards are used, the parameters of the drift corrections can be calculated easily by the regression methods described in Chapter 4. This approach has the advantage that any problems in the drift correction can be more easily detected. For example, the drift correction should be linear. As a consequence all measured intensities, old and new, as shown in Figure 7.1, should be on a straight line. If this is not the case, there is a problem.

4 Uncertainties of Drift Corrections

It is obvious that drift corrections, though very convenient and time saving compared to redoing the whole calibration, are linked to an increase in the uncertainties in the final compositions.

From Equation 7.1, the uncertainty of the corrected intensity is given by:

$$\sigma(\bar{I}_{corr}) = \sqrt{\left(a_1^2\sigma_{\mu 2}^2\left(\bar{I}\right) + \bar{I}^2\sigma^2\left(a_1\right) + \sigma^2\left(a_0\right) + 2\bar{I}u(a_1, a_0)\right)} \qquad (7.3)$$

The uncertainty of the corrected intensity depends not only on the uncertainty of the uncorrected intensity, but also on the uncertainty of the drift correction factors.

Uncertainty of Drift Correction Factors

The simplest way to determine drift correction factors is to measure two different samples with high and low intensities for the element line to be measured. The slope and intercept are given by Equation 7.2. Using the rules for uncertainty propagation, the uncertainties of the drift factors can be calculated 'easily' if the four intensity measurements (two from the calibration and two from the drift correction process) are considered to be independent, *i.e.* not correlated, then:

$$\sigma^2\left(a_0\right) = \left(\frac{I_1'}{I_h' - I_1'}\right)^2 \sigma^2\left(I_h\right) + \left(\frac{I_h'}{I_h' - I_1'}\right)^2 \sigma^2\left(I_1\right) + \left[\frac{I_1 I_1'}{\left(I_h' - I_1'\right)^2}\right]^2 \sigma^2\left(I_h'\right)$$

$$+ \left[\frac{I_h I_h'}{\left(I_h' - I_1'\right)^2}\right]^2 \sigma^2\left(I_1'\right) \qquad (7.4)$$

and

$$\sigma^2\left(a_1\right) = \left(\frac{1}{I_h' - I_1'}\right)^2 \left[\sigma^2\left(I_h\right) + \sigma^2\left(I_1\right)\right] + \left[\frac{I_h - I_1}{\left(I_h' - I_1'\right)^2}\right]^2 \left[\sigma^2\left(I_h'\right) + \sigma^2\left(I_1'\right)\right]$$

$$\qquad (7.5)$$

Note that terms involving $\sigma^2(I_h)$ and $\sigma^2(I_1)$ should be omitted in analysis as they are already included in the calibration.

When multiple-point drift correction is used, the comments on uncertainties in parameters calculated by linear regression in Chapter 4 are applicable. It is important to note that adding more data points to the regression reduces the uncertainty in a regression parameter. From Equation 4.34 we can see that the uncertainty decreases roughly as $1/\sqrt{v}$, where v is the number of degrees of freedom, *i.e.* the number of data points minus the number of fit parameters, which is appropriate when the uncertainties in the data points are constant. As mentioned earlier, in many cases more than two samples are used for drift correcting a method, so that they can be used for all spectral lines simultaneously to reduce the uncertainty in the drift correction without causing additional effort.

Some instruments allow for the replacement of drift correction standards without performing a new calibration. In this case two or more drift corrections are performed consecutively, with the old and new recal samples. The uncertainties do accumulate in this case!

5 Check When Required

Drift corrections are necessary when the changes in the intensities for the elements are significant; *i.e.* when they influence the analytical results significantly. The analytical results are affected significantly when the measured compositions for a sample of known composition differ significantly from their specified or certified compositions. They are significantly different when they are outside the ranges expected from the uncertainties in the measured composition and the uncertainties in the certified values.

There can be no general rule on when or how often to check whether drift correction is required. It depends very much on the number and type of samples being analysed and how often the particular analytical method is being used. It is also dependent on how much an instrument drifts over time and the factors that may be external to the instrument such as argon quality, temperature, *etc.* The ISO standard 10012 gives a clear recommendation of what to do: the analyst must be sure that the analysis is within specification.[1]

Drift correction should often be performed at the beginning of an instrument's life, perhaps every day, or each time the method is used, to establish a clear knowledge of the importance of any drifts. By plotting the changes in the recalibration coefficients over time, one will see how stable the method and the instrument are over a long term and can judge how often drift correction is required. Once this knowledge is acquired, the frequency of drift correction can be adapted and hopefully reduced.

How much drift is acceptable before drift correction is required depends on the expected accuracy of the final results. Whenever there is doubt, an adequately checked sample should be measured to ensure that the instrument is working within specification. The drift correction samples themselves could be used for checking whether a drift correction is needed. Otherwise, a sample of known composition, preferably near the middle of the composition range, could be used. If the change in intensities is greater than expected from their uncertainties, or the difference in composition is greater than expected, a drift correction is warranted. Strictly, to make this decision, we could use statistical decision theory, but pragmatically if the variation is more than two standard deviations or the variation is greater than can be tolerated by the customer, then it is better and safer to do a drift correction.

Any events that might influence the drift behaviour of the instrument significantly, such as cleaning the transmission optics, should be noted in a logbook, to make sure that the instrument is checked for drift before being used for analysis.

6 Sequence for Drift Correction

The following steps are recommended for drift correction for bulk analysis:

1. run drift correction samples
2. check correction coefficients
3. check by analysing a known sample, *e.g.* a CRM

Reference

1. Norme Internationale ISO 10012, Assurance de la qualité des équipements de mesure, Pt. 1, 1992 and Pt. 2, 1997 [French/English].

CHAPTER 8

Bulk Analysis

Glow discharge optical emission spectroscopy (GDOES) was first developed by Werner Grimm for the direct spectrochemical analysis of solid, homogeneous metallic samples. In the early years, many researchers worked on bulk analysis using GDOES especially for the analysis of non-ferrous materials, notably brass and precious metals. Spark Source OES quickly overtook GDOES, at least in terms of speed, cost and ease of analysis.[1] Since then much effort has been put into further improving the performance of Spark spectrometers, now being sold in large numbers. Today it is obvious that Spark is a suitable method for the direct analysis of many metals, especially low-alloy metals. Techniques based on X-ray fluorescence are also applied with much success to the analysis of the main constituents of complex materials including high-alloy metals. Despite the competition from these other techniques, bulk analysis is still an important and necessary area for GDOES. An ability to perform accurate bulk analysis is an essential skill as ever-increasing quality assurance demands are placed on the analyst.

Typical cases where GDOES is used for bulk analysis are:

1. A laboratory using GDOES for surface and interface analysis will, in many cases, find it profitable to use the same instrument for bulk analysis.
2. RF-GDOES is capable of analysing non-conducting materials. Though the performance of GDOES must still be improved for this application, the technique may well prove itself suitable for the routine analysis of homogeneous non-conducting materials.
3. For certain metals with a low melting point, such as lead, tin and zinc alloys, Spark OES has some serious drawbacks linked to the spark discharges, which cause very high plasma temperatures, sample surface melting and material deposition on the electrode.
4. The multi-matrix capability of GDOES allows the analyst an option to use 'sputtering rate corrected compositions' when calibrating the instrument. This can be a special advantage when suitable calibration standards for a given material are sparse.

Bulk analysis is also a good means for judging the general performance of a GD instrument. When determining the composition of a layer or an interface with GDOES, the quality of the result cannot be better than results obtained for a bulk analysis when homogeneous samples and long integration times are used.

1   Precision and Number of Replicates

To optimise the accuracy of an analysis, the intensity measurement should be repeated until the effect of the uncertainty in the mean intensity is less than the effect of the uncertainty in the calibration parameters. Repeating the intensity measurement beyond this will not improve the final result.

Prediction Limits

When it comes to using a calibration curve for analysis, we must now include the statistics of the analysis, namely we need to include the number of analysis measurements (repetitions), m, on each sample, and the uncertainty in their mean. The uncertainty limits of the calibration then become prediction limits of the analysis. For a linear analytical curve without interference terms:

$$\Delta C = t\sigma\,(C) = t \left\{ \frac{\dfrac{1}{w_m m} + \qquad\qquad\qquad\text{(analysis step)}}{s_{y/x}\left[\dfrac{1}{\sum w_i} + \dfrac{(I - \bar{I})^2}{\sum w_i (I_i - \bar{I})^2} \right] \quad\text{(calibration)}} \right\}^{1/2} \tag{8.1}$$

where w_m is the weight for the m analytical measurements. Unlike the weights for the calibration, w_i, which include the uncertainties in the composition of the reference materials as well as those in the intensities, the weight for the analysis step includes only the uncertainty in intensity. This uncertainty in intensity is reflected in the composition axis, as described in Chapter 4.

To prevent this additional source of uncertainty from dominating the overall uncertainty and hence the overall quality of the analysis, the measurement should be repeated several times, *i.e.* $m > 1$. An example is presented in Figure 8.1 for $m = 3$ for our brass calibration.

Detection Limit

The detection limit, c_D, is the smallest elemental composition that can be detected using the smallest detectable signal. It can be determined in many different ways.[2] One particularly useful way is to determine the detection limit from the regression curve itself.[3] Improvements to the regression, e.g. by adding additional data points, by improved precision or by a better regression model, will then be seen to improve the detection limit.

The detection limit can be calculated from Equation 4.45 as the uncertainty in composition estimated for an intensity equal to the background intensity I_B, giving:

$$c_D = t \left\{ \frac{1}{w_m m} + s_{y/x} \left[\frac{1}{\sum w_i} + \frac{(I_B - \bar{I})^2}{\sum w_i (I_i - \bar{I})^2} \right] \right\}^{1/2} \tag{8.2}$$

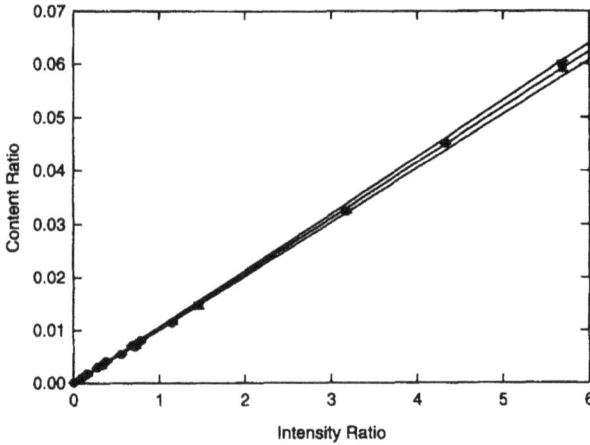

Figure 8.1   *Prediction limits for Al in brass for three measurements*

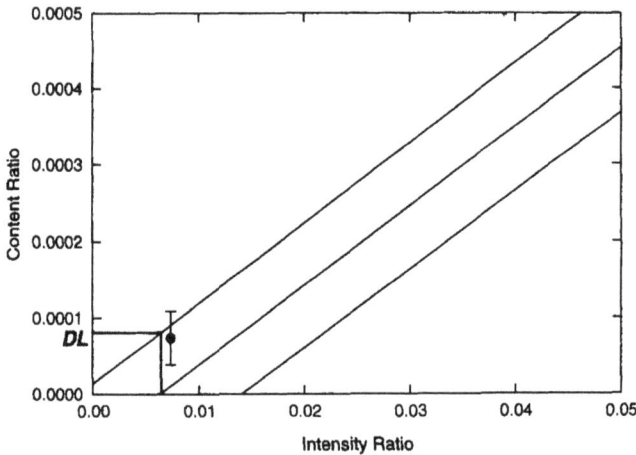

Figure 8.2   *Detection limit for Al in brass calibration based on prediction limit*

where we have assumed m measurements of the intensity. The weight w_m is the inverse square of the standard deviation of the background, s_B, expressed as content, *i.e.* $s_B k_i$, where k_i is the slope. The detection limit will improve with more measurements.

An example using this approach to estimate the detection limit is shown in Figure 8.2, where it is assumed that $m = 3$. It shows the *DL* for Al in brass is 0.000 08 mass % relative to Cu. Since brass is typically 70% Cu, the absolute *DL* is about 0.0056 mass % (56 ppm). Detection limits calculated this way are commonly much larger than those calculated using the SBR-RSDB approach described in Chapter 12.

If we have a large number of samples in the calibration, especially many samples with high weights, then the second and third terms in Equation 8.2 become small, *i.e.*

the uncertainties in the calibration become negligible, and the equation simplifies to:

$$c_D \approx t \left\{ \frac{1}{w_m m} \right\}^{1/2} = \frac{t \, s_B \, k_i}{m^{1/2}} \tag{8.3}$$

To improve c_D we need to increase the number of measurements, m; this will also reduce t. Alternatively we could make instrumental changes to reduce the noise, expressed by s_B, or improve the sensitivity, *i.e.* reduce k_i. Applying Equation 8.3 to the Al brass calibration we get a DL of 0.0009 mass % (9 ppm), which is much lower than the 0.0056 mass % obtained when the uncertainty in the calibration was included.

2 Estimating Uncertainties in Analysis

Different calibration models are used for GDOES. The most general formula for an analytical function is:

$$c_i = f([a], [I]) \tag{8.4}$$

where c_i is the composition of element i, $[a]$ is a set of calibration parameters and $[I]$ is the set of measured intensities, *i.e.* the elemental composition is expressed as some function of the measured intensities and a set of calibration parameters.

Relative calibration modes, e.g. relative compositions as a function of relative intensities, are not significantly different, since the final result is normally expressed as absolute compositions. An additional calculation step is all that is needed.

We will discuss the different calibration functions together with their associated uncertainties. Such uncertainty analysis is an important factor in the choice of the best calibration function.

Direct Calibrations

Direct calibrations, as understood here, are calibration functions using absolute intensities and compositions. The compositions are expressed directly as a function of the measured intensities without ratioing to an internal standard or normalisation. They are the most basic calibration functions used for GDOES.

Linear Calibration Function Without Interference

The simplest case and one often met in GDOES is a linear analytical function without spectral interference:

$$c_i = a_0 + a_1 I_i \tag{8.5}$$

The propagation of uncertainties linked to the parameters a_0, a_1 and the intensity I_i will lead to an uncertainty in the analysed composition, expressed as a variance, of:

$$\sigma^2(c_i) = a_1^2 \sigma^2(I_i) \qquad \text{(analysis step)} \tag{8.6}$$
$$+ \sigma^2(a_0) + I_i^2 \sigma^2(a_1) + 2 I_i \, u(a_0, a_1) \quad \text{(calibration)}$$

where the terms σ^2 are the variances of the different calibration parameters $[a]$ and $[I]$, and the term $u(a_0, a_1)$ is the covariance between the two calibration parameters. The intensity I_i is not correlated with the calibration parameters $[a]$, so the associated covariance between $[I]$ and $[a]$ is zero.

Orders of Polynomials

But in GDOES not all lines are linear, nor are they always free of spectral interference. Usually, the non-linearity is modelled by a power series in the intensities. To understand the influence of non-linearity on the final result, it is again interesting to look at the propagation of uncertainties.

Compared to the linear case, the situation is not significantly different, the formulas are just longer. As an example we will consider intensity as a second-order polynomial without interference:

$$c_i = a_0 + a_1 I_i + a_2 I_i^2 \tag{8.7}$$

The propagation of uncertainties linked to the parameters a_0, a_1 and a_2 and the intensity I_i will lead to an uncertainty, or better variance, in the composition given by:

$$\sigma^2(c_i)c = \underline{(a_1 + 2a_2 I_i)^2 \sigma^2(I_i)} + \sigma^2(a_0) + I_i^2 \sigma^2(a_1) + I_i^4 \sigma^2(a_2) \tag{8.8}$$
$$+ 2I_i\, u(a_0, a_1) + 2I_i^2 u(a_0, a_2) + 2I_i^3 u(a_1, a_2)$$

where the terms σ^2 are the variances of the different calibration parameters $[a]$ and $[I]$, and the terms $u(a_i, a_j)$ are the covariances between any two calibration parameters. The intensity I_i is not correlated with the calibration parameters $[a]$; the associated covariances are therefore zero. The underlined part represents the contribution of the analysis step to the final accumulated uncertainty; the rest is the contribution from the calibration.

The number of covariance terms is larger now; this indicates that we need to take more care that the set of data points for the regression is 'good' enough to ensure that the correlations do not become too large. We also note that the uncertainties are dependent on higher orders of intensity, indicating that precision will decrease for higher elemental compositions and intensities.

Interference Terms

Occasionally, a spectral line from another element is so close to the measured line from the element of interest that it interferes with it (an additive interference). The measured intensity is then a *combination* of the intensities from the two emission lines. The severity of the interference will generally be worse with lower spectral resolution in the spectrometer and with higher composition of the interfering element in the sample. Known spectral interferences for lines commonly used in GDOES are listed in Appendix A.

So we will consider the case of a linear calibration function with one additive interference term:

$$c_i = a_0 + a_1 I_i + b_1 I_b \tag{8.9}$$

where I_b is the intensity of an emission line of the interfering element, scaled by b_1 to match the actual interference. The uncertainty, as a variance, in the content is given by:

$$\sigma^2(c_i) = \sigma^2(a_0) + I_i^2\sigma^2(a_1) + a_1^2\sigma^2(I_i) + b_1^2\sigma^2(I_b) + I_b^2\sigma^2(b_1) \tag{8.10}$$
$$+ 2I_i\,u(a_0, a_1) + 2I_b^2 u(a_0, b_1) + 2I_i\,I_b\,u(a_1, b_1) + \underline{2a_1 b_1 u(I_i, I_b)}$$

where the terms σ^2 are the variances of the different calibration parameters $[a]$ and $[I]$, and the terms $u(a_i, a_j)$ are the covariances between two calibration parameters. The intensities of the two elements I_i and I_b may well be correlated, so their covariance should not be neglected. The intensity I is not correlated with the calibration parameters $[a]$ and $[b]$, so the associated covariances are zero. The underlined parts again represent the contribution of the analysis step to the accumulated uncertainties.

One of the purposes for showing these lengthy formulas for calculating the uncertainties in the measured compositions is to make GDOES users aware of the implications of using higher order polynomials and interference terms. If they are required and sufficient data points are included in the regression to give reliable values for the additional regression parameters, then they can greatly enhance the resulting accuracy compared with inadequate linear functions without corrections. If they are not required, however, or if the regression data are inadequate to define them reliably, then they may improve the beauty or 'look' of the calibration curves, as more points will fall on the calibration line, but in effect they may seriously degrade the accuracy of the final result.

Relative Calibration Modes

Relative calibration functions express relative compositions as a function of relative intensities:

$$(c_i/c_{\text{maj}}) = f([a]; [I/I_{\text{maj}}]) \tag{8.11}$$

where $[a]$ are the calibration parameters. All compositions and intensities are expressed relative to the composition of a major element in the matrix material, *i.e.* Cu or Zn in brass, for example.

The content of the major element expressed in a mass fraction can be calculated as:

$$c_{\text{maj}} = \frac{1}{1 + \sum c_i/c_{\text{maj}}} \tag{8.12}$$

The absolute composition of each element is then given by:

$$c_i = (c_i/c_{\text{maj}})\,c_{\text{maj}} = \frac{(c_i/c_{\text{maj}})}{1 + \sum c_j/c_{\text{maj}}} \tag{8.13}$$

This analytical mode is particularly interesting when:

- deviations in repeated intensity measurements of the analysed element are correlated with those of the major element; this improves the precision
- (pragmatically) the regression is better than with other modes; this improves accuracy

The first condition usually indicates that the excitation conditions are not perfectly stable. The second condition, more frequently met, is due to varying sputtering rates or other matrix effects. If the sputtering rate varies from one calibration sample to the next, this will affect all elements at the same time and by the same amount. The ratio of two intensities and the ratio of two compositions in the sample should therefore not be affected.

It must, however, be remembered that relative methods only give reasonable results when the analytical response of the reference line, *i.e.* the intensity of the spectral line of the major element, used for calculating the intensity ratio, is linear over the range covered by the calibration curve.

The calculation of uncertainties for relative calibration modes is similar to that for the direct modes, except for the final step of calculating the absolute compositions. Using the uncertainty propagation rules, we find that the uncertainty in the composition of the major element can be expressed as:

$$\sigma^2(c_{maj}) = c_{maj}^4 \sum [\sigma^2(c_i/c_{maj})] \tag{8.14}$$

The uncertainties of the absolute compositions are calculated from the uncertainties of the relative compositions using the uncertainty propagation rule. To simplify the equation, let $x_i = c_i/c_{maj}$. If we consider that all the x_i are independent, *i.e.* neglecting covariance, we can write:

$$\sigma^2(c_i) = \sum_{j=1}^{n} \left[\frac{dc_i}{dx_j}\right]^2 \sigma^2(x_j) \tag{8.15}$$

Calculating the derivatives using the relationship:

$$\frac{d\left(\dfrac{1}{a+bx}\right)}{dx} = \frac{-b}{(a+bx)^2} \tag{8.16}$$

we obtain:

$$\sigma^2(c_i) = \left[\frac{1+\sum_{j\neq i} x_j}{\left(1+\sum_{j} x_j\right)^2}\right]^2 \sigma^2(x_i) + \left[\frac{x_i}{\left(1+\sum_{j} x_j\right)^2}\right]^2 \sum_{k\neq i} \sigma^2(x_k) \tag{8.17}$$

This calibration mode is very interesting for GDOES as sputtering rates do change significantly when the chemical composition of a sample changes. However, from an analysis of the uncertainty propagation, we find that the uncertainty in one elemental composition depends on the uncertainties of all the other elements. This implies that if one element is incorrectly determined, then all others will be adversely affected as well.

3 Homogeneity

The remarks made on homogeneity of calibration and drift correction materials are fully applicable to analysed samples, in the sense that non-homogeneous samples will lead to variations in the analytical results. Unlike reference materials, which can be selected carefully to give optimum results, production samples intended for analysis generally must be analysed as they are.

In this case, the small amount of material analysed by GDOES may present a disadvantage when employing the GDOES technique for the bulk analysis of industrial samples, since a number of measurements on different areas of the sample will be necessary to give a good average composition. On the other hand, GDOES may be used successfully to investigate and verify the homogeneity of such non-ideal materials.

Sample preparation for bulk analysis should be the same as for calibration and drift correction materials.

4 Accuracy

Precision indicates the level of uncertainty in a measurement result. Accuracy, on the other hand, is an estimate of how well this measurement result agrees with the 'true' value. But in chemical analysis, the true composition of a sample, even if it is certified, is ironically not known. So, instead of the 'true' value, we use the 'accepted' value for estimating the accuracy. Hence for certified materials we use the 'certified' value. Hopefully, the accepted value is close to the true value, otherwise we will have a poor estimate of accuracy and may well bias our results away from the true value. It is for this reason that modern certified reference materials are measured with a variety of different techniques to avoid bias from one particular technique or analytical laboratory.

The results for the analysis of brass L6 are shown in Table 8.1. The uncertainties shown are 95% confidence limits on the mean of three measurements. The uncertainty in the certified values is based on Equations 4.27 and 4.28, since no uncertainties were given on the certificate, while the uncertainty in the measured values is based on our brass calibration and Equations 8.14 and 8.17.

5 Grade Analysis

In many industrial applications, the exact composition of a sample is not required; rather it matters whether the composition of certain key elements falls within predefined limits. This is called 'grade' analysis. Besides the specific grades defined within individual laboratories, there are thousands of internationally and nationally defined grades or classes of materials, and often these are common to a particular industry. Iron (metal), for example, is defined as Fe (element) containing $2.0\% \leq C \leq 4.5\%$.

Table 8.1 *Bulk analysis (mass %) of brass CTIF L6*

Element	Line (nm)	Certified	Uncertainty	Measured	Uncertainty	Accuracy
Cu	224.700	66.55	0.3	66.0	1.4	−0.6
Zn	334.502	30.26	0.3	30.8	1.0	0.6
Si	288.158	1.25	0.03	1.24	0.04	−0.01
Ni	341.477	1.21	0.03	1.18	0.02	−0.03
Sn	317.505	0.250	0.010	0.257	0.014	0.007
Pb	220.353	0.205	0.009	0.201	0.005	−0.004
Al	396.152	0.139	0.007	0.144	0.004	0.005
Fe	371.994	0.085	0.006	0.089	0.004	0.004
Mn	257.610	0.055	0.004	0.060	0.002	0.005

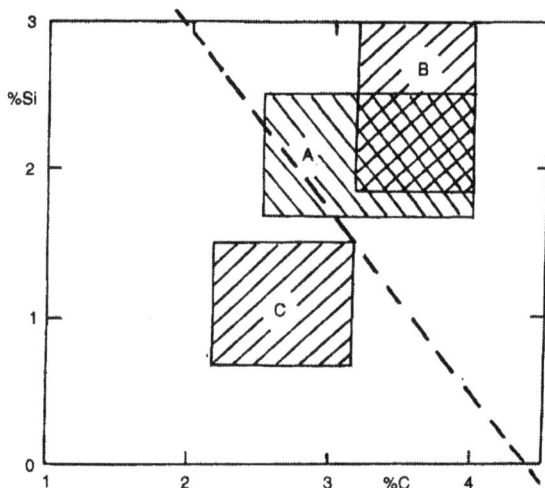

Figure 8.3 *Typical C and Si compositions in graphitic cast iron: A, grey; B, ductile and C, malleable; the dashed line indicates the approximate eutectic composition[4]* (Reproduced with permission from Z. Weiss, *Spectrochim. Acta* B, 1996, **51**, 863)

Part of a grade specification for graphitic cast iron is shown graphically in Figure 8.3. Usually, it is a simple pass or fail test: if the analysed element is inside the range it passes the grade, if outside the range it fails.

An example of the chemical composition (grade) for one type of white cast iron is shown in Table 8.2. As can be seen, the composition range for C in this grade is quite restricted. This is because, in this steel, abrasion resistance and repeated shock resistance are inversely related to the C composition.

6   Presentation of Results

Bulk analysis is normally conducted for a customer who is principally interested in the elemental compositions. A typical result form is shown in Table 8.3, and a typical grade analysis form is shown in Table 8.4.

Table 8.2 *Chemical composition for white cast iron, AS 2027, Ni Cr4-600[5]*

Element	Min %	Max %
C	2.8	3.2
Si	1.5	2.2
Mn	0.2	0.8
P		0.15
S		0.12
Cr	8.0	10.0
Mo		0.5
Ni	4.0	5.5
Fe	Remainder	

Table 8.3 *Typical form for bulk analysis results*

GDOES Bulk Analysis Report		
Report Number	2002 AMA 211	
Sample name	L6-1	
Sample preparation	Lathe	
Operator's name	Nelis-Payling	
Method	Brass-0	
Anode size (if varied)		
Number of samples	1	
Number of burns	3	
Element	*Content (mass %)*	*SD mean*
Cu	66.5	0.51
Zn	30.3	0.31
Si	1.25	0.25
Ni	1.21	0.21
Sn	0.25	0.05
Pb	0.21	0.03

If the results are to be published in scientific journals, the following additional information is often required:

- instrument details (manufacturer, model number, etc., including any in-house modifications)
- spectrometer type (polychromator or monochromator, PMT or solid state)
- wavelengths of spectral lines used
- GD source (type, *e.g.* Grimm-type or Marcus-type; RF or DC; continuous or pulsed; constant power or constant current, *etc.*)
- source conditions (*e.g.* pressure and power, or current and voltage)
- calibration samples (name and manufacturer/distributor)

Table 8.4 *Typical form for grade analysis*

GDOES Grade Analysis Report			
Report Number	2002 AMA 210		
Sample name	L6-1		
Grade	Special Brass		
Sample preparation	Lathe		
Operator's name	Nelis-Payling		
Method	Brass-0		
Anode size (if varied)			
Number of samples	1		
Number of burns	3		
Element	*Content (mass %)*	*Pass ($\checkmark$ or $\times$)*	*Accepted rage*
Cu	66.5	$\checkmark$	60–70
Zn	30.3	$\checkmark$	25–35
Si	1.25	$\times$	0.5–1.0
Ni	1.21	$\checkmark$	1–2
Sn	0.25	$\checkmark$	<1
Pb	0.21	$\checkmark$	<1

7 Case Study: Cast Iron

The bulk analysis of cast iron is an interesting example for GDOES. In this analysis, it is important to distinguish two different families of cast irons: (i) white or chill-cast iron and (ii) grey or as-cast iron. While white cast iron is very homogeneous, grey cast iron contains graphite particles of various sizes and shapes. The free graphite particles pose a serious problem when an attempt is made to analyse the material with Spark source discharges.

The graphite particles are relatively difficult to sputter; as a result we can see the effect of differential sputtering in the depth profiles, as demonstrated in Figure 8.4. Before the C composition stabilises, more than 20 μm needs to be sputtered through first. This typically requires a pre-burn time of more than 300 s.

Despite the differences in the degree of differential sputtering between white cast and grey cast iron, white cast iron samples can be used for calibrating the instrument, and grey cast iron samples can still be analysed accurately with GDOES, provided the pre-burn time is sufficient. This can be seen in Figure 8.5 where white (various series, including CKD and BAS) and two grey (BS 20G and BS 20K) cast irons are displayed on the same calibration curve.

The accuracy that can be attained with GDOES for the analysis of cast iron is shown in the work of Winchester and Miller, presented here in Figure 8.6. Comparison with Figure 4.8 shows that it is comparable with the accuracy obtained with certified reference materials; for example, approximately 1% relative at 1 mass %.

Some of the limitations of GDOES for the analysis of as-cast irons must be mentioned. Using a 4 mm anode the analysis area is about 12 mm^2. If the sample is not homogeneous in this area, the analytical results will not be reproducible. In particular, when large graphite particles are present, the analysis should be repeated on different spots. Another possible problem when analysing bulk samples with GDOES is the

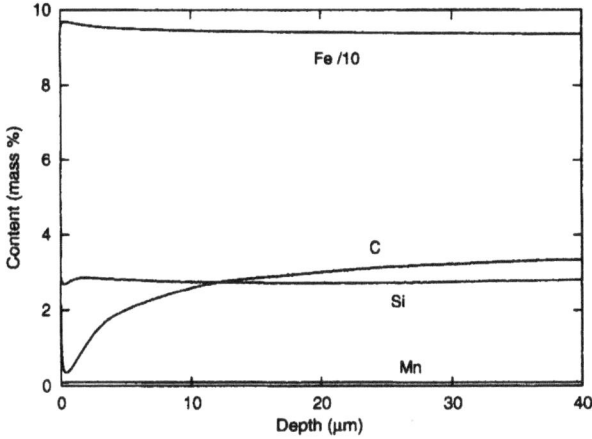

Figure 8.4 *Compositional depth profile of an industrial grey cast iron sample*

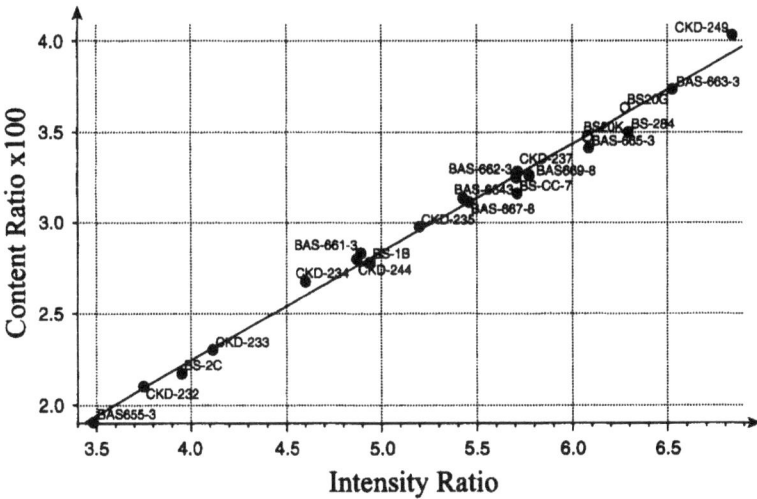

Figure 8.5 *Calibration for C in cast iron, ratioed to Fe, using white (•) and grey (○) cast samples*

potential for air to be trapped in cavities within the sample. They will alter the plasma and completely destroy the quality of the analysis. This holds true for other analytical techniques as well.

8 Sequence for Bulk Analysis

The following steps are recommended for the bulk analysis of solid samples:

1. select the appropriate method
2. prepare the samples
3. check if drift correction is required

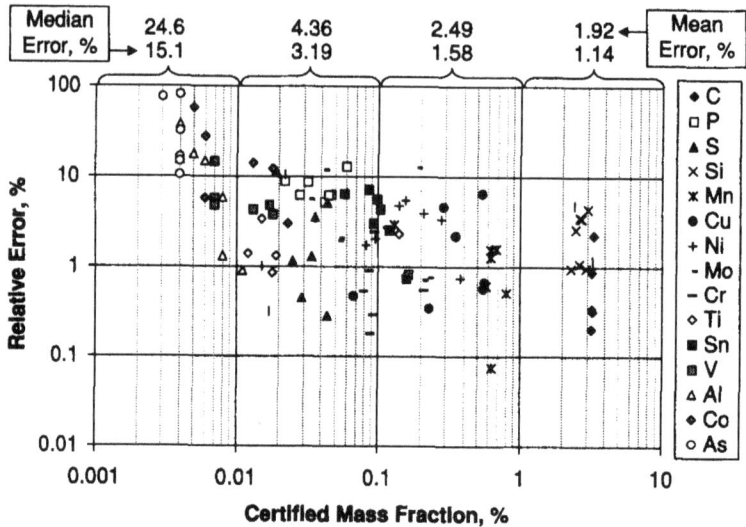

Figure 8.6 *Relative errors (accuracy) for 15 analytes in grey cast iron*[6]
(Reproduced with permission from M. Winchester and J.K. Miller, *J. Anal. Atom. Spectrom.*, 2001, **16**, 122)

4. perform drift correction (if required)
5. analyse the samples (including repeated measurements)
6. check that the values are sensible
7. check uncertainties (beaware of the precision)
8. check grade (if required)
9. report the results

References

1. K.A. Slickers, Die automatische Atom-Emissions-Spektralanalyse, Buchvertrieb K.A. Slickers, Giessen, FRG, 1992.
2. P.W.J.M. Boumans, in *Glow Discharge Optical Emission Spectrometry*, R. Payling, D.G. Jones and A. Bengtson (eds), John Wiley & Sons, Chichester, 1997, 440–51.
3. D.L. Massart, B.G.M. Vandeginste, L.M.C. Buydens, S. De Jong, P.J. Lewi and J. Smeyers-Verbeke, *Handbook of Chemometrics and Qualimetrics: Part A*, Elsevier, Amsterdam, 1997, 195, 432–5.
4. Z. Weiss, *Spectrochim. Acta B*, 1996, **51**, 863.
5. Australian Standard AS 2027, Standards Australia, 2002.
6. M. Winchester and J.K. Miller, *J. Anal. Atom. Spectrom.*, 2001, **16**, 122.

CHAPTER 9

Qualitative Depth Profiling

Our intention here is not to provide a review of GDOES depth profiling applications. These can be seen in other publications such as those listed in Further Reading section at the end of this book. Our intention, rather, is to choose examples that illustrate typical situations in depth profiling for the analyst to demonstrate some of the strengths of the technique and some ways to avoid its weaknesses.

Compositional depth profiles, discussed in the next chapter, although having the advantage of providing quantitative numbers directly, cannot be better than the qualitative depth profiles used to produce them.

Qualitative results are a very efficient resource for process engineering and defect analysis. To ensure this, it is essential that there be an ongoing discussion between the GD analyst and the process or product engineer. Both will learn much from the discussion; outsourcing by the engineer to a GD analytical laboratory without this close relationship will not be as beneficial. Here is a situation where GD specialists can encourage their customers to discuss their analytical needs, rather than simply ask for a depth profile: the outcome can be surprisingly good.

1 Near Surface

Surface studies with GDOES typically vary from the first few nanometres down to some tens of nanometres in depth. Since typical sputtering rates are 50–100 nm s^{-1}, it takes only about 30 ms to go through the first 2 nm and about 0.15 s for the first 10 nm. Some analysts think that it would be better to go slower through this region to improve depth information. But if nothing else is changed, other than the sputtering rate (e.g. by reducing the average power), then the total amount of information does not change. The measured intensities are essentially proportional to the number of sputtered atoms, so simply varying how quickly these atoms leave the surface will not change the total measured intensity.

This is illustrated in Figure 9.1 for the near surface of galvanised steel. The coating is 99.5% Zn, but the surface is rich in Al (as a mixed oxidehydroxide). The as-measured intensities are shown in Figure 9.1a for 30 W (No. 1) and 17 W (No. 2) applied power. The difference in applied power means that No. 2 has about half the sputtering rate of No. 1. The lower sputtering rate means that the Al peak occurs later

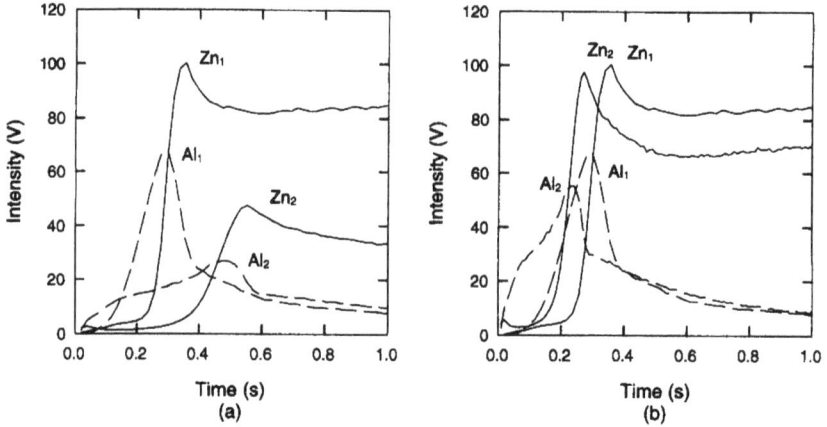

Figure 9.1 *Surface of galvanised steel at two different RF applied powers, No. 1 at 30 W, No. 2 at 17 W: (a) as-measured, (b) with time for No. 2 × 17/30 and intensities for No. 2 × 30/17*

and with lower intensity. If the time and intensities for No. 2 are scaled by the ratio of the two powers (*i.e.* by 30/17), then the result is as shown in Figure 9.1b. Clearly, there are some apparent differences in shape, due to differences in the way the plasma started in the two measurements, but the information content is essentially the same. If we integrate the total Al intensity in Figure 9.1a, in each case from time 0 to just past the Al peak, to determine the surface Al, we obtain the same number 570 Vs, indicating that, as expected, the amount of surface Al has not changed with applied power.

The integrated intensity remains constant. The peak intensity reduces as the sputtering rate is reduced and the time of measurement increases. Making a rough estimation we obtain:

$$I \propto \frac{1}{t}; \qquad \text{for noise } N \propto \frac{1}{\sqrt{t}} \tag{9.1}$$

The signal-to-noise ratio therefore increases with the inverse of the square root of time. Certainly, this kind of rough estimation needs to be looked at with some scepticism, but it tells us that reducing the sputtering rate and increasing the total measurement time do not necessarily improve the signal quality. Indeed, faster sputtering means better signal-to-noise, so slower sputtering may make things worse. Improvements can only be made by varying the plasma conditions to optimise plasma startup and depth resolution.

The main concern with surface studies is to have a sufficient number of points to see any surface layers and to have low noise so that any changes are clear. These two requirements are at odds with each other: to get more points in the depth profile we need to reduce the signal averaging time, while to reduce noise we should increase the signal averaging time. Obviously, the choice for the operator is a compromise. To see a layer clearly it is good to have at least 5–10 points, hence to see a layer in the first 2 nm, an averaging time of 5 ms should be sufficient, and since deeper layers are generally thicker it should rarely be necessary to have an averaging time less than

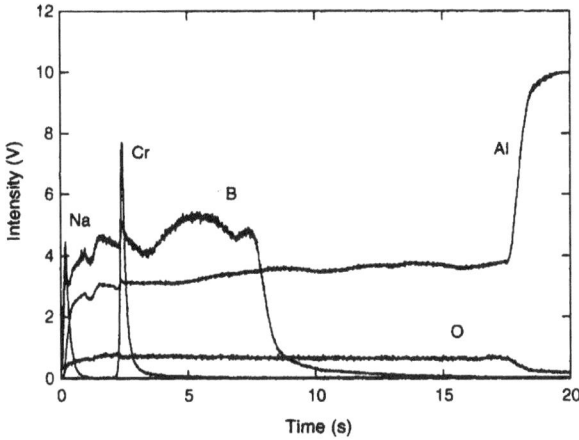

Figure 9.2 *Surface of a sequential anodised aluminium, showing an oxide film approximately 360 nm thick and a boron-rich outer layer, spiked with Cr from a Na_2CrO_4 solution at a depth of about 47 nm*[1]

5 ms. In many cases, 10 ms or 0.01 s is sufficient, and shows much less noise. An example of a surface depth profile is shown in Figure 9.2.

Presentation of too many data points in a depth profile, on the other hand, is also to be avoided. It is not much better than presenting too many digits; they are not significant! Generally, a few hundred points per element are sufficient to see any trends.

An alternative to increasing the averaging time is to collect the data very quickly and then use some form of sophisticated smoothing, such as fast Fourier transform (FFT), time window or polynomial (spline) fitting. These smoothing algorithms use special numerical filters in either the time or frequency domain. Which type of smoothing works best will depend on the type of noise and the number of data points included in the smoothing; see, for example, the surface of galvanised steel, in Figure 9.3, where both signal averaging and FFT smoothing appear to work well.

Smoothing should be used for eliminating non-relevant information and maintaining the relevant information. The smoothing function should therefore be defined in terms of the expected, specified depth resolution after smoothing. Care should be taken when smoothing non-equally spaced data points, as many algorithms are written for equally spaced data. But this is only a software problem.[2,3]

An important question often posed in surface studies is: is there more or less of an element in one sample compared with another? Provided the signals from the element are much above their backgrounds so that their depth profiles can be seen distinctly, the qualitative depth profile is usually sufficient to answer this question. For similar materials, analysed under the same source conditions and if linear spectral lines are used, the intensities are proportional to the number of sputtered atoms. So, if we integrate the intensities within the layer of interest, see Figure 9.4, then we have effectively counted the total number of atoms of the element in that layer, *i.e.* the coating weight of the element, except that without calibration we do not know the

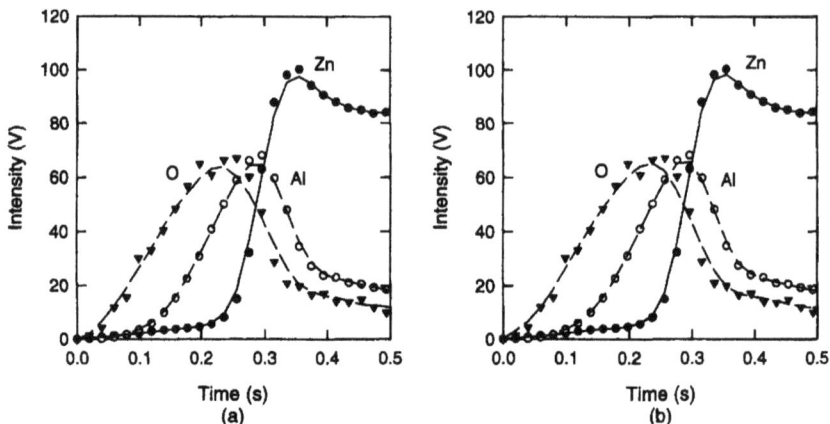

Figure 9.3   *Near surface of galvanized steel: points are raw data, solid lines are (a) signal averaged and (b) FFT smoothed*

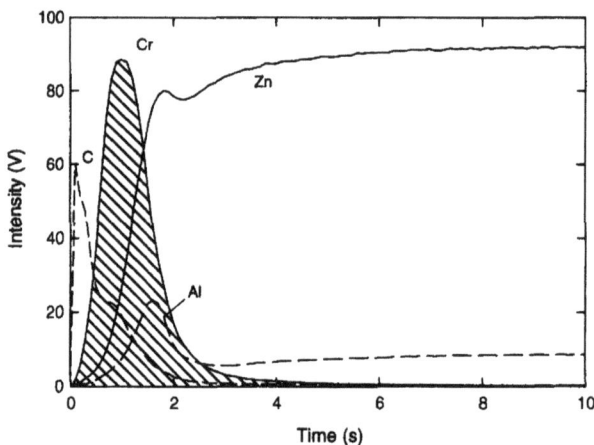

Figure 9.4   *Chromate-treated galvanized steel surface, showing integral (area) under the Cr qualitative depth profile, which should vary linearly with the coating mass of surface Cr*

proportionality constant. So, if we then ratio the values for an element from two similar samples, the ratio is the relative composition of the element in the layer. Also, if we can compare these values with quantitative results from some other technique (*e.g.* ICP-OES), then we have an alternate way for calibrating the results from our qualitative depth profiles, by plotting known coating weight of an element versus its integrated intensity.

While the quantification of depth profiles (discussed in Chapter 10) can remove some artefacts, especially those associated with changing sputtering rates, it introduces additional uncertainties from the calibration. Qualitatively, the difference in intensities may be significant for an element in two similar samples, but after quantification of the difference in the estimated compositions may no longer appear significant. One must ask the right question. Looking at different lines of the same element may indicate why this difference occurred and help avoid traps.

Source Conditions

When performing depth profile analysis with GDOES, qualitative or quantitative, unless one is studying surface cleanliness, the sample surface should be properly cleaned and all grease should be removed. As discussed later in Chapter 11, hydrogen in the plasma has a strong effect on the emission yield of most spectral lines, making a simple interpretation of the observed intensities difficult. When thin layers are to be analysed, it is often difficult to distinguish between effects due to pollution of the sample surface and the discharge chamber from 'real' effects on the sample surface.

The measured intensities of elements such as C, H and O tend to decrease exponentially in depth profiles when they are due to pollution. It is often interesting to compare the observed intensities with the results obtained on a freshly polished sample. Silicon wafers, because of their high purity and cleanliness, are good test pieces to assess the effect of residual gases on the measured intensity profiles. Generally, long flush times (several minutes) are required to minimise pollution in the source prior to surface studies.

Glow discharge is a fast analytical tool. When thin surface layers are analysed, the sputtering rates of a few micrometer per minute are rather high. It is therefore important to set the conditions of plasma ignition to allow fast stabilisation of the plasma. This is particularly true for parameters such as the setting of the impedance matching system for a fixed frequency RF discharge or the pressure in the discharge chamber. These parameters usually use mechanical parts for regulation and optimisation. The stabilisation time may be long compared to the analytical requirements. It is usually necessary to run several trials to find the optimum conditions.

2 Coatings

Qualitative depth profiling is a rapid means for studying coatings. In many cases, for example, because of the small number of samples in the study, or the absence of suitable reference materials, it is not feasible to take the time to do a calibration to convert the results into compositional depth profiles. In these cases the qualitative results must suffice.

Typical questions that can be answered by qualitative depth profiles of coatings are:

1. Is a particular layer present?
2. How do the thicknesses of a coating compare on different, similar samples? or how do the coating weights of a layer compare on different, similar samples?
3. Is the coating homogeneous?
4. Is an element present throughout the coating?
5. Is there more or less of an element in a coating on different, similar samples?

Qualitative depth profiles must be interpreted with care as they are subject to several artefacts, some of which can be removed in compositional depth profiles, but not all. It is good to discuss the interpretation with a process engineer or the producer of the sample; GDOES should be considered as part of the total process and not an independent tool.

The main factors affecting the results are:

- varying sputtering rates in going through different layers
- varying plasma conditions in different layers affecting intensities, *e.g.* varying voltage if in constant power or constant current mode
- varying H content in different layers affecting intensities
- varying background signals in different layers
- melting of low-temperature materials
- differential sputtering effects

Because of possible differences in sputtering rates, it is often difficult to be sure of the relative thicknesses of different coatings on a single sample. It is better to compare the thicknesses of similar layers on different samples. Because of variations in plasma conditions, it can be difficult to compare small changes in an element in one layer with another quite different layer on the same sample. Again, it is better to compare intensities for similar layers on different samples.

If you see large differences in an element between different samples, check to see that there is no large difference in H in these samples, as H is known to cause large changes in intensities, increasing some lines and decreasing others. Before interpreting qualitative data in detail, check the variations of free plasma parameters: pressure in the constant voltage, constant current mode; voltage in the constant pressure, constant power mode; or the impedance matching system in RF. Any significant variation may cause a change in emission yield or sputtering rate.

It is difficult to be sure what is happening to trace levels of elements in depth profiles, because the background signals can change from one layer to the next; see, for example, Figure 6.11.

The melting or pyrolysing of low-temperature materials can produce unstable depth profiles, *i.e.* large variations in intensities in homogeneous coatings, and large variations from sample to sample. If it is suspected that this is happening, one should increase the cooling or run the sample at a lower power until one gets stable, reproducible depth profiles.

Differential sputtering effects are caused, typically, by differences in the sputtering rates of different grains in the sample, or by the development of sputter cones around inclusions. They can usually be seen in a very rough crater bottom. They reduce the depth resolution and should not be confused with real effects in the coatings, such as diffusion.

Source Conditions

Most coatings analysis with GDOES does not require any special sample preparation, other than ensuring that the sample is suitable for mounting on the source. Oily samples, however, should be degreased, unless, of course, the oil is the aim of the study.

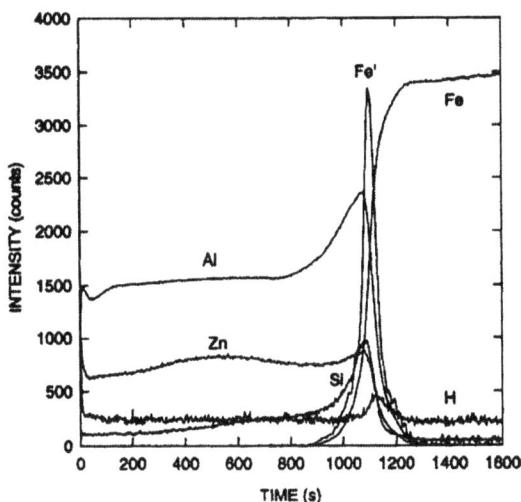

Figure 9.5 *Qualitative depth profile of Galvalume: Al–Zn–Si-coated steel. The differentiated iron signal, Fe', is a way to indicate the interface position and width*[4]
(Reproduced with permission from R. Payling, in *Glow Discharge Optical Emission Spectrometry*, R. Payling, D.G. Jones and A. Bengtson (eds), John Wiley & Sons, Chichester, 1997, 589–95)

Choose source conditions, *e.g.* power or current and voltage, suitable for the materials, reducing the power if the materials are heat sensitive. Adjust the source conditions, *e.g.* pressure or voltage, to optimise the depth resolution in the region of the coating of most interest. This is generally not in the substrate. Select a signal averaging time, which is not too short otherwise the file size will be too big, and not too long otherwise there will not be enough points to see the features of interest. Most software allows the selection of different signal averaging times for different time segments, so, for example, you can select a very short averaging time at the beginning of the depth profile to capture rapid changes at the surface and longer averaging times later in the depth profile to reduce noise and data size.

3 Interfaces

The artefacts in qualitative depth profiles are most noticeable at the interfaces between layers. An example is shown in Figure 9.5, where the increase in Al signal approaching the steel substrate is mainly due to an increase in sputtering rate. The Al content, we know from compositional depth profiles and from other techniques, should in fact decrease there. The Galvalume shown is a 55% Al, 43% Zn and 1.5% Si coating on steel, produced by the continuous hot-dipping of steel through a molten metal bath. Normal galvanised steel, on the other hand, has a much lower Al content, typically 0.1–0.5% Al. For galvanised steel an increase in Al content is also seen at the interface, see Figure 9.6, but in this case it is real, due to the formation of a Zn–Al–Fe ternary alloy layer between the coating and steel. So a detailed knowledge of the sample helps in the interpretation of elemental depth profiles.

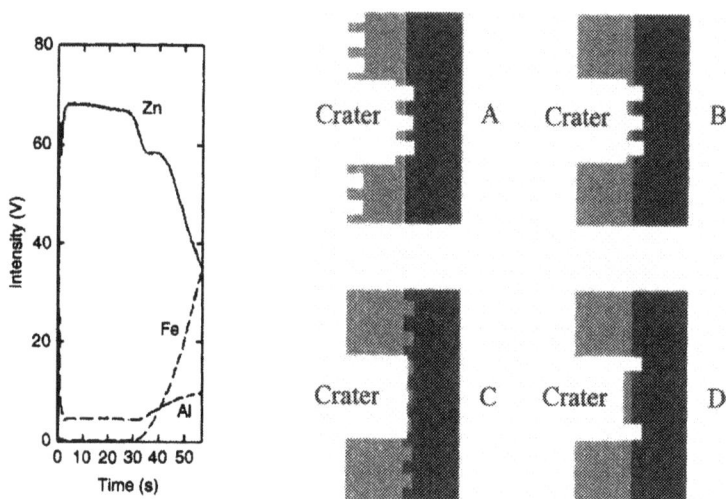

Figure 9.6 *Factors affecting depth resolution: A, rough surface; B, differential sputtering due to microstructure; C, rough interface and D, crater shape. The qualitative depth profile shown is from galvanized steel, stopped at the interface between the coating and steel substrate*

Probably the main requirement at interfaces is high depth resolution to see variations in composition including diffusion. But the depth resolution of interfaces in GDOES is limited by the following:

- initial surface roughness
- differential sputtering in outer layers
- roughness of the interface of interest
- crater shape

Which effect dominates will depend largely on the nature of the samples themselves. The reasons for interface broadening are illustrated in Figure 9.6.[4]

The first two contributions, surface roughness and differential sputtering in superficial layers, can be reduced by polishing the sample or by removing outer layers with solvents or acid etching, taking care not to alter the interface of interest. For the Galvalume sample shown in Figure 9.5, a dramatic improvement can be seen at the interface between the Al–Zn–Si coating and the steel substrate after acid etching to remove the free coating, as shown in Figure 9.7.

When the limit is caused by the roughness of the interface of interest, there is little that can be done except to be aware that the apparently wide interface is due to roughness and not due to a thick interface. The classic example of this is galvanneal, a Zn–Fe–coated steel; this will be considered in more detail in Chapter 10.

When the limiting effect is crater shaped, significant improvements can be achieved by optimising the source conditions for the interface and not for the coating or substrates as is often done. You can check this either by stopping the sputtering at the interface and examining the shape of the crater there, or by simply comparing

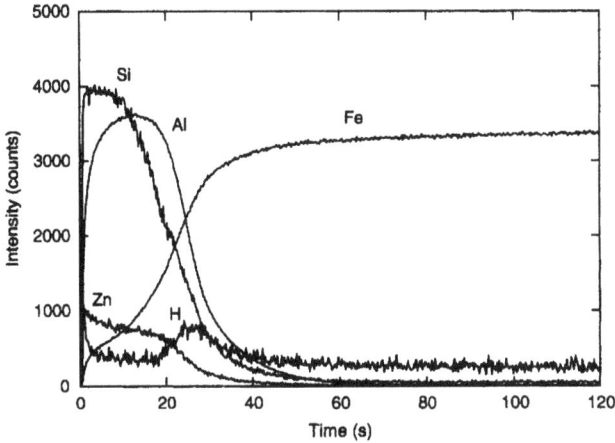

Figure 9.7 *Qualitative depth profile of Galvalume that has been acid etched to remove the coating but with the Fe–Al–Zn–Si quaternary alloy interface left intact on the steel*

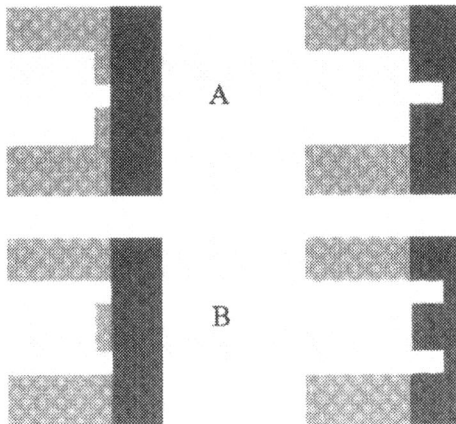

Figure 9.8 *Effect of crater shape at the interface of a slow sputtering material (light grey) on a fast sputtering substrate (dark), where A represents a convex crater shape and B represents a concave shape*

depth profiles recorded with varying plasma conditions and choosing the conditions that best show the interface.

It is occasionally observed that the shapes of intensity profiles at an interface are non-symmetric. This can be explained easily by the combination of a non-flat crater and a significant difference in sputtering rates between the two layers. Imagine a non-flat crater encountering a perfectly defined interface. For our purpose, consider that the non-flat crater bottom can be approximated in two dimensions by a square, with a nozzle at the centre to represent a convex crater, or nozzle at either side to represent a concave crater, see Figure 9.8. When the inner layer (nozzle) reaches the interface we begin to see signals from the substrate. If the substrate sputters faster than the coating (*e.g.* paint on steel), the nozzle will then speed up while the remainder of the crater is still in the coating. The depth resolution will be poor at the interface, and signals from the outer layer will be observed for a long time. When the substrate sputters slower than the coating, the depth resolution will improve at the interface. In fact, the finally observed intensity time profile is a convolution of the different effects, leading to the apparent broadening of the interface. A detailed study of this effect can be found in Präßler *et al.*[5]

4   Comparison of Intensity Profiles

An important means of displaying differences in qualitative depth profiles is by over-laying elemental profiles. A simple example is shown in Figure 9.9, illustrating the differences in coating thickness between two Zn-coated steel samples.

In another example, see Figure 9.10, a quick examination of light and dark-coloured regions on a PVD-coated flat steel spring shows a difference in composition and a penetration of the coating by the substrate (indicated by Fe).

In a third example, H ions were implanted in a silicon wafer.[6] The implant compo-sition was known by the process engineer to be 5×10^{16} ions cm^{-2}. The silicon wafers were annealed at different temperatures. The GDOES depth profiles, in Figure 9.11, immediately show the effect of annealing on the distribution of the implanted ions. This is a typical example where qualitative depth profiles and a close collaboration between the process engineer and the analyst can lead to quick answers. Calibration of the GD instrument is not required here, as the process engineer knows how many ions are implanted in the sample. The total analysis time for each sample was only a few minutes.

There are many other examples in material science, or the semiconductor industry, where the comparison of qualitative results may be sufficient in the control and

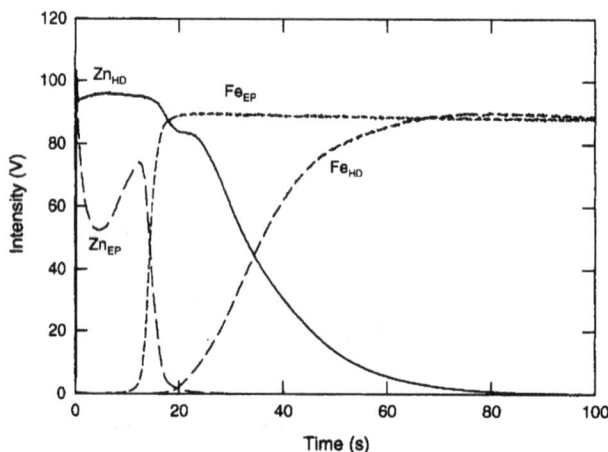

Figure 9.9 *Comparison of coating thickness for two different Zn-coated steel samples: (light) electroplated, (dark) hot-dipped galvanized*

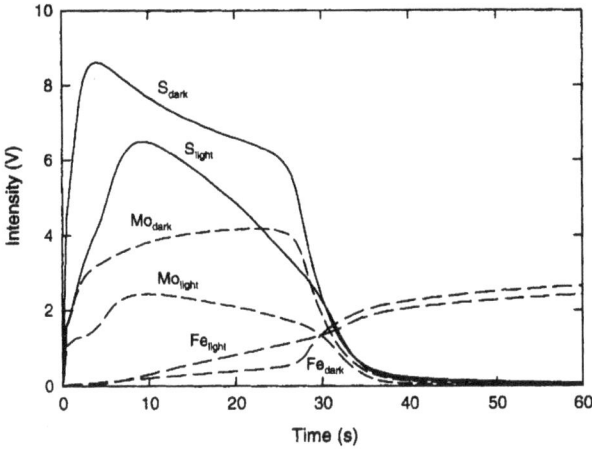

Figure 9.10 *Comparison of two regions on a PVD-coated flat spring: (solid) light area, (dashed) dark area*

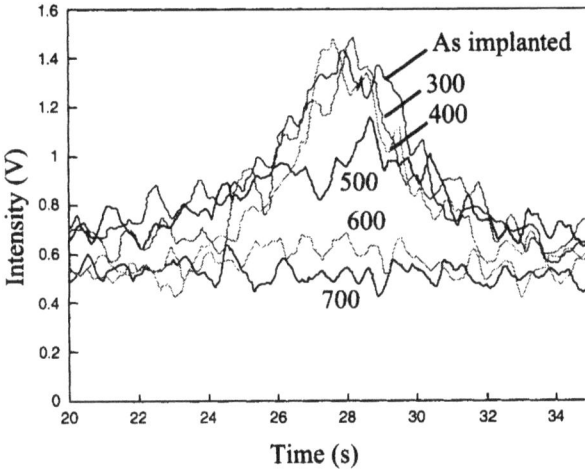

Figure 9.11 *Comparison of qualitative hydrogen intensities. H^+ ions implanted 10^{16} ions/cm^2. Annealing temperature given in degrees Celsius*
(Reproduced with permission from application Report GDS3878Y, Rigaku Co., 2000)

development of coating processes. In CVD processes, for example, it is fairly easy to check with GDOES for the presence of an element such as Cl in a layer, or for other defects directly linked to the process.

5 Presentation of Results

Qualitative depth profiling is normally conducted for customers who are principally interested in understanding the nature of their sample. A typical results form is shown in Table 9.1.

Table 9.1   *Typical form for qualitative depth profiling*
            results

GDOES Qualitative Depth Profiling Report

Report number   2002 AMA 212
Sample name   AuNiP-Brass1
Sample description   Au on NiP on Brass
Sample preparation   None
Operator's name   Nelis-Payling
Method   H-S-2
Anode size (if varied)
Element   Au, Ni, P, Cu, Zn, Pb
Comment   Thin gold layer
Graph of qualitative depth profile

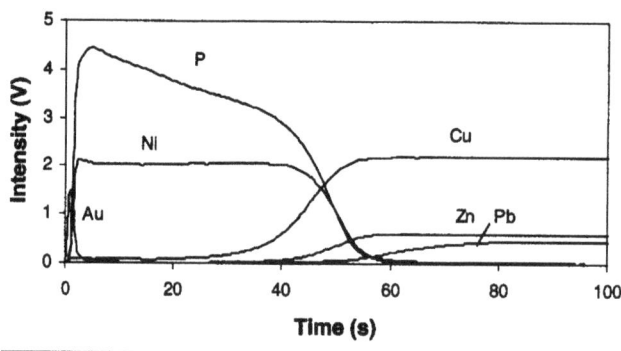

If the results are to be published in scientific journals, the following additional information is often required:

- instrument details (manufacturer, model number, *etc.*, including any in-house modifications)
- spectrometer type (polychromator or monochromator, PMT or solid state)
- wavelengths of lines used
- GD source (type, *e.g.* Grimm-type or Marcus-type; RF or DC; continuous or pulsed; constant power or constant current, *etc.*)
- source conditions (*e.g.* pressure and power, or current and voltage)

6   Sequence for Qualitative Depth Profiling

The following steps are recommended for qualitative depth profiling:

1. determine appropriate method
2. prepare samples (if necessary)
3. choose a suitable total sputtering time, sufficient to see everything of interest

4. choose a suitable averaging time(s), which is a compromise between number of points, noise and file size
5. run depth profile
6. examine the depth profile for stability, if necessary reduce power and repeat
7. compare with related depth profiles
8. report results

References

1. K. Shimizu, G.M. Brown, H. Habazaki, K. Kobayashi, P. Skeldon, G.E. Thompson and G.C. Wood, *Surf. Interface Anal.*, 1999, **27**, 24.
2. M. Labarrere, J.P. Krief and B. Gimonet; *Le filtrage et ses applications*, Cepadues–Editions, Toulouse, France, 1988.
3. P.A. Lynn and W. Fuerst, *Introductory Digital Signal Processing*, 2nd edn, John Wiley & Sons, Chichester, UK, 1994.
4. R. Payling, in *Glow Discharge Optical Emission Spectrometry*, R. Payling, D.G. Jones and A. Bengtson (eds), John Wiley & Sons, Chichester, 1997, 589–95.
5. Frank Präßler, V. Hoffmann, J. Schumann and K. Wetzig, *Fresenius J. Anal. Chem.*, 1996, **355**, 840.
6. Application Report GDS3878Y, Rigaku Co., 2000.

CHAPTER 10
Compositional Depth Profiling

Quantification of intensity versus time plots is achieved by applying the calibration curves to the measured intensities. In fact, the quantification procedure basically consists in applying the calibration function to the data point by point. In this sense quantification is just a repetition of bulk analysis, except for the shorter integration times and the fact that measurements are usually not repeated several times on different spots. The quality of the calibration is therefore crucial to the quality of the compositional depth profile (CDP).

Calibration for CDP was discussed in Chapter 8, while a more theoretical description will be given in Chapter 11. We will concentrate here therefore on more analytical aspects, especially the uncertainty estimation, interpretation and representation of CDP analyses.

As described in Chapter 6, the calibration function for CDP relates sputtering rate corrected compositions $c_i q$ to measured intensities, $i.e.$:

$$c_i q = f([a], [I]) \tag{10.1}$$

where $[a]$ is a set of regression parameters and $[I]$ are the measured intensities. Assuming that we have measured all major elements present in the sample, we can normalise these values to 1, or 100%. This allows us first to calculate the compositions rather than just the sputtering rate corrected ones:

$$c_i = c_i q / \sum_j c_j q \tag{10.2}$$

and then to calculate the relative sputtering rate ($i.e.$ the relative sputtered mass per second) during the period when the intensities were measured:

$$q = q \sum c_i = \sum c_i q \tag{10.3}$$

We can then apply a model for estimating the density of the analysed material to convert the information on sputtering rates to sputtered depth. This is described in more detail in Chapter 11. Finally, we are able to display the typical CDP graph of elemental compositions as a function of depth. Several examples will be shown later in this chapter.

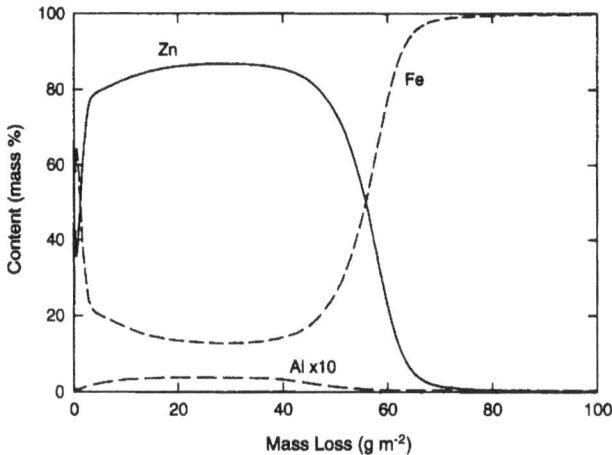

Figure 10.1 *Double layer Zn–Fe coating on steel, plotted as content versus sputtered mass/surface. This allows direct observation of coating masses*

It should be mentioned, at this point, that the sputtered mass, directly related to the coating weight, is measured directly from the sputtering rates without making use of a specific model on the density of the analysed material, whereas the information on depth requires such a density model calculation. It is therefore interesting to express GD results in terms of sputtered mass or coating weight, whenever possible, to avoid making use of possibly inaccurate models on density, see, for example, Figure 10.1.

1 Mass %

CDP results are typically represented as graphs of chemical composition in mass % as a function of sputtering depth. The outer layers are presented first, to the left of the graph; inner layers are to the right at increasing depth. When discussing this kind of graph with a process engineer, the GD specialist should be aware that process engineers often think the other way round. The first layer is the one that has been applied first; it is the inner layer of the coating. This may seem an unnecessary comment, but it is a common error when production people first look at CDP results.

2 Atomic %

GDOES found its first applications in the metal processing industries, which mostly use mass % or coating weight to express the chemical composition of an alloy or coating (probably because they are used to mixing tonnes of material A with tonnes of material B). Other commercial sectors, such as the semiconductor industry or paint industry, often express chemical composition in atomic % or atomic numbers per volume, comparing numbers of atoms or molecules (take a few billion Si atoms, add some B atoms and one has doped silicon).

Table 10.1   *Comparison of mass % and atomic % for some common materials*

Material	Element	Mass %	Atomic %
Al 12Si	Al	88.0	88.4
	Si	12.0	11.6
Al$_2$O$_3$	Al	52.9	40.0
	O	47.1	60.0
Brass	Al	0.14	0.33
	Cu	69.9	70.3
	Zn	30.0	29.3
Cast iron	C	3.0	12.4
	Si	2.0	3.5
	Fe	95.0	84.1
Pigmented	H	4.6	42.8
Polymer[1]	C	42.2	33.2
	O	32.0	18.9
Stainless steel	Cr	18.0	19.2
	Ni	12.0	11.3
	Fe	70.0	69.5
TiN	N	22.6	50.0
	Ti	77.4	50.0

GDOES is capable of expressing analytical results either way. There is no significant difference in accuracy in expressing the analytical results either way. It is therefore recommended to satisfy the needs of the people requesting the results.

It is possible to perform a CDP calibration directly in atomic %, but most commercial software does not offer this possibility. The decision to plot results in mass % or atomic % is therefore taken during the presentation of the analysis results. To convert between mass %, c_i, and atomic %, n_i, the following equations are used:

$$n_i = \frac{c_i}{w_i} \frac{100}{\sum_j \frac{c_j}{w_j}} \quad \text{or} \quad c_i = \frac{100\, n_i w_i}{\sum_j n_j w_j} \tag{10.4}$$

where w_i is the atomic weight of element i. The conversions between mass % and atomic % are shown in Table 10.1 for some common materials. The larger the differences in atomic weights between elements in the material, the larger the effect of conversion.

Expressing results in atomic % is particularly sensible when chemical bonds are predominant in the analysed material. When analysing a TiN layer, for example, you would expect 50% Ti and 50% N expressed in atomic %, otherwise the TiN layer would not be stoichiometric. Similarly, expressing an oxide layer in atomic % helps to assign the oxidation state, when the compositions of possible stable oxides are known. A stoichiometric SiO$_2$ layer would be one-third Si and two-thirds O, in atomic %.

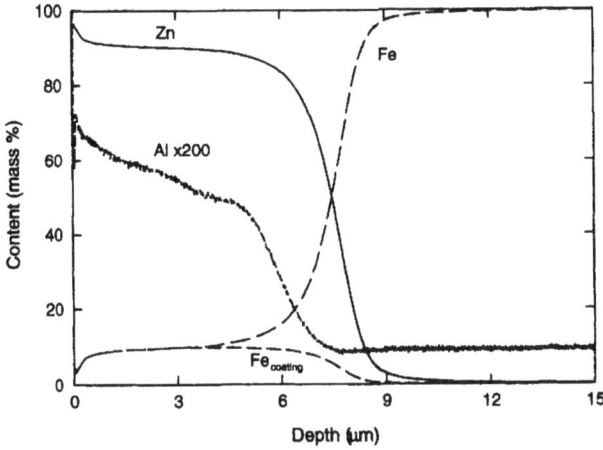

Figure 10.2 *Galvanneal, a Zn–Fe-coated steel, showing Fe in the steel substrate (Fe) separated from Fe in the coating (Fe$_{coating}$)*

3 Coating Weight

Coating weight is another measure for describing the amount of coating material on a substrate. It is usually expressed in mass per unit surface (*e.g.* g m^{-2}). It is widely used in the steel and car-manufacturing industries. Standard methods for measuring coating weights are ICP-OES and X-ray-based techniques, with GDOES offering some advantages over these, especially in analysis time and the additional information contained in the depth profiles.

To calculate the coating weight from a CDP analysis we simply sum the sputtered mass over the depth of the coating:

$$m_i = q_{Ref} \sum_{j=0}^{N} (c_i q_M / q_{Ref})_j \, \Delta t_j \tag{10.5}$$

where N is the number of measurement intervals needed to sputter through the coating. The sputtered area is included in the definition of the reference sputtering rate, q_{Ref} (see Chapter 11 for more details). Assumptions on the density of the sputtered material do not enter the calculation; they therefore do not cause any trouble. The difficult task lies in defining the coating itself, *i.e.* where precisely it starts and where it stops.

Let us look in detail at the CDP analysis of the Zn–Fe (galvanneal) coating on steel shown in Figure 10.2. It is an important application for GDOES. Galvanneal is made by passing (hot-dipping) steel coil through a molten metal bath, mostly containing Zn. After dipping the coil is kept hot for some seconds to allow Fe from the steel to diffuse into the Zn layer. The result is a Zn–Fe alloy coating whose properties depend critically on the crystal structure of the alloy, which in turn depends critically on the amount of Fe in the coating.

Looking from the steel outwards, the Fe composition should start near 100% in the steel, show a rapid decrease to the coating composition, usually between 9 and 13 mass %, and then decrease again at the outer surface of the coating. Examination of the coating/substrate interface with microscopic techniques, such as optical microscopy or scanning electron microscopy, shows that the interface between the Zn–Fe layer and the steel base is narrow, well defined but rough. The interface roughness broadens the observed elemental profiles in GDOES.

When analysing the interface with GDOES it is easy to determine the total amount of Zn by integrating over the whole Zn depth profile, since there should be no Zn in the steel. But for Fe, it is not quite so obvious where the Fe in the coating ends and the Fe in the steel begins, *i.e.* how much of the Fe depth profile belongs to the coating and how much to the steel.

When the composition and coating weights of this kind of material are analysed by ICP, the Zn–Fe layer is dissolved in inhibited acid. Dissolution stops when the Fe in the solution reaches a predetermined value. The amounts of Zn and Fe in the coating are then determined by analysing the acid. Though there is a continuing discussion as to how much, if any, of the steel is dissolved, the values obtained are reproducible and process engineers are used to these values. Any determinations of coating weights in GDOES are therefore guided by this experience with ICP measurements and are biased towards obtaining the same results. It is a nice example of how things can be easy and difficult at the same time.

One solution in GDOES is to recognise that the apparently large interface between the coating and the steel is due mainly to the roughness of the interface.[2,3] Throughout the interface, the Fe in the coating should therefore follow the same trend as the Zn profile in the interface. In this way we can separate $Fe_{coating}$ in the coating from Fe in the steel. The result is shown in Figure 10.2. Such a method appears to give similar results to ICP analysis, and would be valid for similar profiles in other materials.

4   Uncertainty of the Calculated Depth in CDP

In Chapter 6 we showed how to calculate the uncertainties in composition for bulk analysis. The methods described there can be readily extended to calculate the uncertainties in composition in CDPs. In this chapter we will, instead, outline how to estimate the uncertainty in the calculated depth.

For each time interval Δt, the sputtered depth Δz is:

$$\Delta z = N_s\, q_{Ref} \frac{\Delta t}{\rho} \tag{10.6}$$

where N_s is a normalisation factor given by:

$$N_s \equiv \sum c_i q \tag{10.7}$$

and the density ρ is given by:

$$\frac{1}{\rho} = \sum \frac{c_i}{\rho_i} = \frac{1}{\sum c_j q} \sum \frac{c_i q}{\rho_i} \tag{10.8}$$

From the definitions for the normalisation factor and the density we can simplify Equation 10.6 to:

$$\Delta z = q_{Ref} \Delta t \sum \frac{c_i q}{\rho_i} \tag{10.9}$$

The uncertainty in sputtered depth can be estimated by uncertainty propagation as:

$$\sigma^2 (\Delta z) = (\Delta t)^2 \left[\sum \frac{c_i q}{\rho_i} \sigma (q_{Ref}) \right]^2 + (q_{Ref} \Delta t)^2 \left[\sum \frac{c_i q}{\rho_i} \sigma (\Delta t) \right]^2$$
$$+ (q_{Ref} \Delta t)^2 \sum \left[\frac{\sigma (c_i q)}{\rho_i} \right]^2 \tag{10.10}$$

Possible uncertainties in the densities of the different elements are neglected here. If we also neglect the uncertainty due to the time increment, then the remaining sources of variance are due to the reference sputtering rate and the $c_i q$ values.

The sputtered depth after N depth measurements is:

$$z = \sum_{k=1}^{N} \Delta z_k \tag{10.11}$$

The uncertainties of each depth measurement will, of course, add. But first we must take a more careful look at the correlations between the different variables, in particular, between the corrected compositions $c_i q$. These are calculated from the intensities and the calibration constants. We can consider two consecutive intensity measurements I^k and I^{k+1} to be independent and not correlated. The calibration constants, on the other hand, will be the same for each, *i.e.* completely correlated. To have a better picture we will develop the calculation of the cumulated depth z further:

$$z = q_{Ref} \sum_k \left[\Delta t_k \sum_i \frac{(c_i q)^k}{\rho_i} \right] \tag{10.12}$$

where k is the index for the measurement step and i is the index for the elements. For simplicity we will assume that the time step is constant and accurate throughout the depth profile and that we have used linear calibration functions for the $c_i q$; hence:

$$z = q_{Ref} \Delta t \sum_k \left[\sum_i \frac{a_i I_i^k + b_i}{\rho_i} \right] \tag{10.13}$$

Even with this simplification it is still difficult to derive a formula for the uncertainty, which does not include correlations between different elements. For the calculation of the uncertainty we consider the variables and their variances to be independent for two consecutive measurement steps, which may or may not be true,

Table 10.2   *CDP results for various Zn–Al coatings*

Coating	Coating thickness (μm)	Uncertainty (μm)
Galvanneal	7.5	1.0
Galvanised	20.1	2.8
Zn–Al–Si	19.8	2.7
Zn 5% Al	7.9	1.1
Zn–Ni	5.9	0.8

to obtain:

$$\sigma^2(z) = \Delta t \left(\sum_k \sigma^2 (q_{Ref}) \left(\sum_i \frac{a_i I_i^k + b_i}{\rho_i} \right)^2 + q_{Ref} \sum_i \sigma^2 (a_i) \left(\sum_k \frac{I_i^k}{\rho_i} \right)^2 \right.$$

$$\left. + q_{Ref} \sum_{i,k} \sigma^2 (I_i^k) \left(\frac{1}{\rho_i} \right)^2 + q_{Ref} \sum_i \sigma^2 (b_i) \left(\sum_k \frac{1}{\rho_i} \right)^2 \right) \qquad (10.14)$$

The part of the uncertainty linked to the calibration curves will certainly add, without avoiding the covariances, because the calibration parameters from one time interval to the next are the same, *i.e.* fully correlated. The part of the uncertainty linked to random noise in the measurement should not be correlated, so again we have to add the variances and not the uncertainties.

It must also be stated that systematic errors, such as those in density, are not taken into account here. If known they should be corrected before calculating the uncertainties; if unknown, they should be estimated and taken into account for the estimation of the different parameters.

The coating thicknesses for some commercial Zn-based coatings, estimated from their CDPs, are shown in Table 10.2, along with their uncertainties calculated by using Equation 10.14 for 95% confidence ($t = 1.96$).

Based on these data, the largest source of uncertainty in depth, by far, is $s(q_{Ref})$, *i.e.* the estimated uncertainty in the reference sputtering rate. Equations 10.13 and 10.14 then simplify to:

$$\frac{s(z)}{z} = \frac{s(q_{Ref})}{q_{Ref}} \qquad (10.15)$$

5   Presentation of Results

Compositional depth profiling is normally conducted for customers who are interested both in understanding the nature of their samples and gaining quantitative values for some key features, such as the average compositions in various layers, coating thickness or coating mass. A typical form of results is shown in Table 10.3.

Table 10.3 *Typical form for compositional depth profiling results*

Report number	2020 AMA 213
Sample name	AuNiP-Brass1
Sample description	Au on NiP on brass
Sample preparation	None
Operator's name	Nelis-Payling
Method	H-S-2
Anode size (if varied)	
Elements	Au, Ni, P, Cu, Zn, Pb
Comments	Thin gold layer
Graph of compositional depth profile	

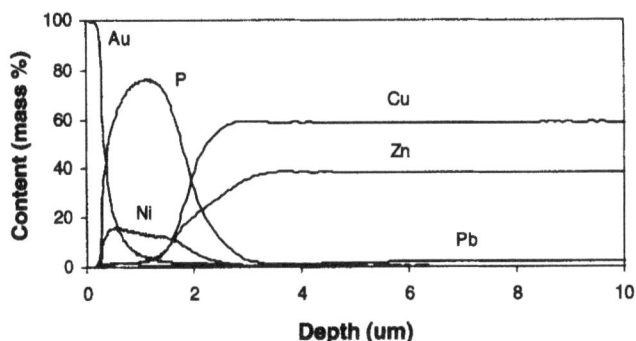

Coating thickness (μmm)	Au 0.37; NiP 1.5
Coating mass (g m^{-2})	Au 6.1; Ni 8.4 P 1.6: NiP 10.0

Coating	Element	Average content (mass %)
Au	Au	98
Ni P	Ni	73
	P	13

Summary

Depth (μm)	Element 1	Element 2	Element 3	Element 4	Element 5
	Au	Ni	P	Cu	Zn
0.5	20.1	59.6	15.8	1.4	1.7
1.0	5.2	75.2	13.7	2.0	2.1
1.5	2.2	69.6	11.9	7.4	7.5
2.0	1.6	31.9	6.0	39.0	20.8
2.5	1.2	11.0	2.0	56.1	29.2
3.0	0.7	2.5	0.4	59.3	36.4
3.5	0.4	0.9	0.1	58.7	38.8
4.0	0.3	0.5	0.006	58.9	38.9

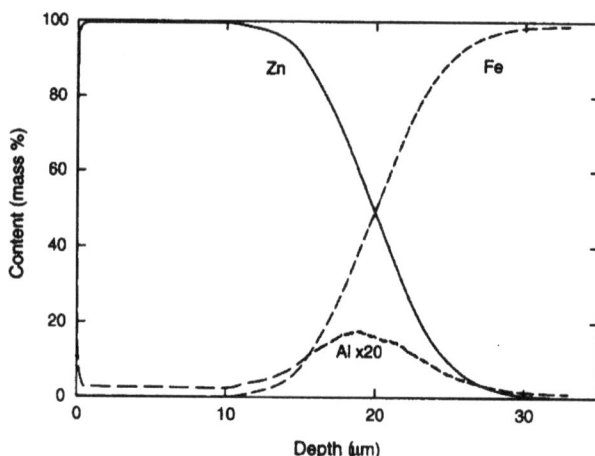

Figure 10.3 *CDP of galvanized steel*

If the results are to be published in scientific journals, following additional information is often required:

- instrument details (manufacturer, model number, *etc.*, including any in-house modifications)
- spectrometer type (polychromator or monochromator, PMT or solid state)
- wavelengths of lines used
- GD source (type, *e.g.* Grimm-type or Marcus-type, RF or DC, continuous or pulsed, constant power or constant current, *etc.*)
- source conditions (*e.g.* pressure and power, or current and voltage)
- calibration samples (name and manufacturer/supplier)

6 Case Study: Galvanised Steel

One of the most important Zn-coated steel products is continuous hot-dipped galvanised steel sheet. A long coil of preheated steel is passed continuously through a molten metal bath containing Zn, a small amount of Al (typically 0.2 mass %) and possibly a small quantity of Pb or Sb. As the steel sheet exits the bath, air jets control the coating thickness by pushing excess molten metal back into the bath.

The calibration for Zn-based coatings was described in detail in Chapter 6. The CDP of galvanized steel using this calibration is shown in Figure 10.3.

A related product is Galvalume, described in Chapter 9, where the molten bath has a composition of Al 55%-Zn 43%-Si 1.5%. A CDP of this product, using the same calibration as for Figure 10.3, is shown in Figure 10.4. The high contents of Al and Zn are not completely soluble in each other, so they tend to form Zn-rich and Al-rich regions in the coating. This non-homogeneity is evident in the CDP.

The CDPs of such coatings are used to measure coating thicknesses and the coating masses of various elements, average coating compositions and for checking the distribution of key elements. A summary of results from two inter-laboratory (round robin) studies is shown in Table 10.4. The SMT round robin was a formal study,

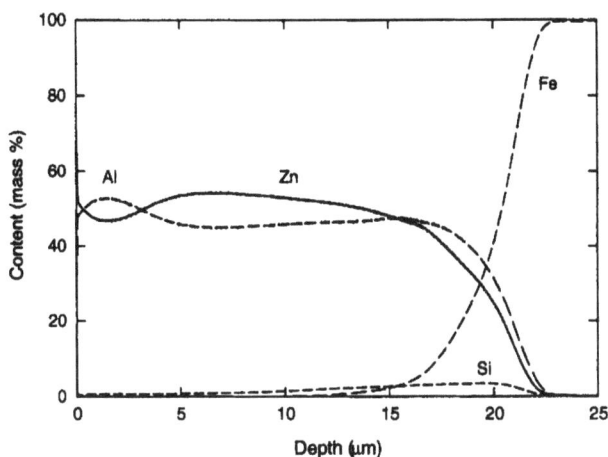

Figure 10.4 *CDP of Galvalume, a Zn–Al–Si-coated steel*

Table 10.4 *Results on Galvalume, a Zn–Al–Si-coated steel, from two inter-laboratory studies*

Analyte	Wet chem.	SMT RR mean[4]	SD	RF RR mean[5]	SD
Total coating mass (g m^{-2})	82.4	82.2	9.0	81.6	7.1
Coating thickness (μm)	21.6	20.9	2.3	20.3	1.5
Zn (mass %)	45.2	44.9	3.4	45.1	2.2
Al (mass %)	53.2	52.5	3.2	53.3	2.4
Si (mass %)	1.6	1.9	0.5	1.6	0.3

being part of the European work to develop a GDOES standard for Zn-based coatings, and involved 5 different coatings and 12 laboratories, 10 DC and 2 RF powered instruments; the RF round robin was an informal study using the same samples but involving five laboratories, all using RF powered instruments. Generally, the two studies showed very similar results.

Inter-laboratory studies tend to find larger standard deviations than are found in results within a single laboratory. As the reasons for the differences between laboratories are identified and the laboratories adopt standard practices, these inter-laboratory differences tend to decrease dramatically.

7 Sequence for CDP Analysis

The following steps are recommended for qualitative depth profiling:

1. determine appropriate method
2. prepare samples (if necessary)

3. check whether drift correction is required
4. choose a suitable total sputtering time, sufficient to see everything of interest
5. choose a suitable averaging time(s), which is a compromise between number of points, noise and file size
6. run depth profile
7. quantify
8. measure parameters, *e.g.* coating mass and coating thickness
9. report results

References

1. D.G. Jones, in *Glow Discharge Optical Emission Spectrometry*, R. Payling, D.G. Jones and A. Bengtson (eds), John Wiley & Sons, Chichester, 1997, 541–6.
2. ISO CD 16962, Surface chemical analysis. Zn and Al-based metallic coatings by GDOES, 2002.
3. K. Kakita, K. Suzuki and S. Suzuki, in *Glow Discharge Plasmas in Analytical Spectroscopies*, R.K. Marcus and J.A.C. Broekaert (eds), John Wiley & Sons, Chichester, 2003, 209–29.
4. A. Bengtson, S. Hägström, E. Lo Piccolo, N. Zacchetti, R. Meilland and H. Hocquaux, SMT-CT96-2080, 1998.
5. R. Payling, *Multinational RF Round-Robin*, unpublished report, 1999.

CHAPTER 11
Theory

The GD source generates the signals we use for analysis. Understanding the theory of this source will help in optimising the conditions used in analysis and in interpreting the results. Here we describe the key aspects of GD theory: the nature of the plasma, of sputtering and the approaches used for quantitative analysis, in particular, compositional depth profiling.

1 Plasma

The argon glow discharge is a low-pressure plasma. It can also be thought of as a weakly ionised gas. It is electrically neutral, which means that the number of negative charges, mainly free electrons, is balanced by an equal number of positive charges, mainly positively charged argon ions. But it is not homogeneous, nor in thermal equilibrium. When sputtered atoms from the sample enter this plasma, there are many different types of collisions (see Figure 11.1), some of whose cross-sections are not known in detail, and difficult to model with particle simulation (*e.g.* Monte Carlo methods) or with fluid models, and the choice between these two remains a little arbitrary.

These properties make it a difficult plasma to understand in detail, and can cause some difficulties in analysis. The variations in plasma density give rise to non-linear calibration curves for some emission lines through self-absorption, and, because the plasma dimensions can change with changing plasma conditions, they produce other non-linear behaviours in emission. The complexity of the plasma makes it difficult to model changes in emission yield, *i.e.* the average number of photons emitted per sputtered sample atom. But despite these difficulties, the power of the technique for quantitative analysis remains.

If we were to look inside the GD source, through the source window to the optical spectrometer during an analysis at typical conditions, we would notice a bright violet–pink glow completely covering the sample surface. For historical reasons these conditions are called an 'abnormal' discharge. It means that as we increase the current the voltage also increases, because the increase in current is achieved by an increase in current density as the entire cathode already takes part in the process. If we could then move inside the plasma to the sample surface we would notice that the space

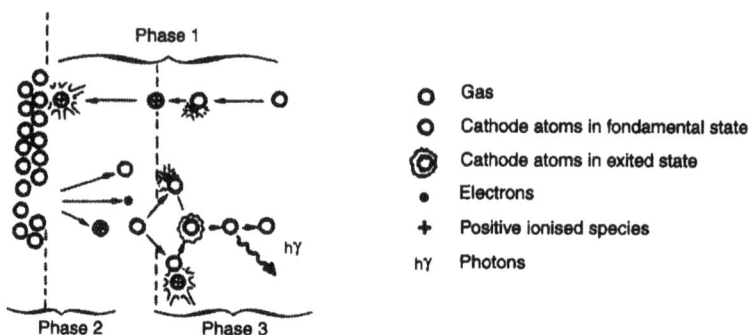

Figure 11.1 *Main processes occurring in the analytical glow discharge.*[1] *In phase 1, argon atoms are ionised and bombard the cathode surface; other possible bombarding species (not shown) are fast argon neutrals and sputtered cathode ions. In phase 2, sputtered species include cathode atoms, ions and electrons. In phase 3, sputtered atoms and ions can undergo various collisions, become excited and then de-excite by emitting a photon*
(Reproduced with permission from M. Bouchacourt and F. Schwoehrer, in *Glow Discharge Optical Emission Spectrometry*, D.G. Jones and A. Bengtson (eds), John Wiley & Sons, Chichester, 1997, 51–3)

immediately in front of the sample is dark (actually it is not completely dark, just much darker). This dark space, also called the cathode fall, extends for about 1 mm from the sample. It is also called the cathode sheath. Most of the potential difference between cathode and anode occurs here. Then as we move back out again, there is an intense bright region, called the negative glow (NG), extending several millimetres, and gradually fading as we move back towards the source window.[2] The NG plasma is nearly field-free and constitutes the bulk of our GD plasma (see Figure 11.2). In some glow discharge sources there is another weaker, bright zone further out, called the positive glow, but in Grimm-type sources the anode is so close to the cathode (sample) that normally no positive glow is present.[3]

The applied voltage causes electrons to flow with high energy from the sample surface towards the anode (see Figure 11.3). The electrons and argon interact to form a plasma. In this plasma, most of the potential is dropped across the cathode dark space (CDS) in front of the sample.[3] The high-energy electrons there collide with argon atoms causing ionisation of the argon. Some sample atoms are also ionised and head back at speed towards the sample. The positive argon ions and sample ions are then driven by the negative bias in the CDS to collide with the sample surface. Along the way the ions suffer other collisions and lose much of their energy.[5] Many of the ions also regain an electron through charge-transfer collisions with neutral atoms, converting a neutral atom into a slow ion and themselves into a fast neutral. They keep most of the energy they had as an ion and their momentum means that they continue towards the sample. As the ions and neutrals strike the surface of the sample, they have sufficient kinetic energy left to cause sputtering of the sample surface.

The sputtered atoms move away from the surface where they suffer many collisions with the argon gas and very rapidly slow down to thermal speeds of around 500 m s.[6]

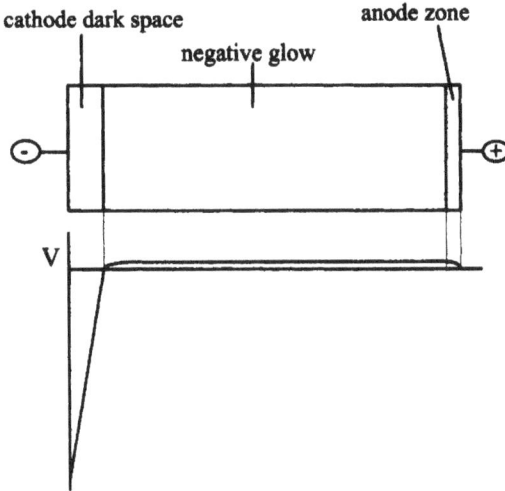

Figure 11.2 *Schematic representation of the analytical glow discharge and the corresponding potential distribution*[4]
(Reproduced with permission from A. Bogaerts and R. Gijbels, *Spectrochim. Acta* B, 1998, **53**, 1)

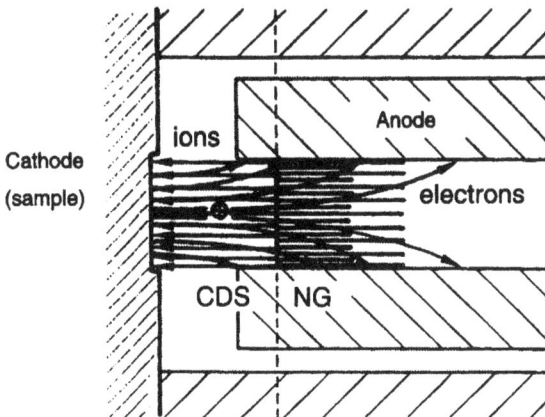

Figure 11.3 *Paths for ions and electrons in the analytical glow discharge.*[1] *Also indicated is the CDS and NG*
(Reproduced with permission from M. Bouchacourt and F. Schwoehrer, in *Glow Discharge Optical Emission Spectrometry*, D.G. Jones and A. Bengtson (eds), John Wiley & Sons, Chichester, 1997, 51–3)

Many are knocked back onto the sample surface. Some are ionised and contribute to sputtering of the sample. At the edge of the CDS and into the NG region, there are plenty of electrons with sufficient energy (greater than about 4 eV) to cause excitation. Argon atoms are also excited in this region into their metastable state and collisions with metastable argon atoms can also cause excitation of the sputtered atoms.[3] Once excited, the sputtered atoms then de-excite by emitting photons with characteristic wavelengths. These are the lines of analytical interest; most of the glow, however, comes from excitation and de-excitation of the argon.

2 Modelling the Plasma

The inhomogeneous and non-equilibrium nature of the 'confined' analytical glow discharge makes it a difficult plasma to model. If it were homogeneous and in equilibrium, we might be able to devise a series of general equations to predict particle distributions and energies, and the populations of excited levels and intensities. These equations would use fluid models for the particle distributions, the Poisson equation for the electric field and Boltzmann's equation for the populations of excited levels. But these would not describe our GD plasma adequately. Alternatively, we could model the behaviour of thousands of individual particles through Monte Carlo simulations and link this to the Poisson equation (the so-called particle-in-a-cell model) to arrive at the average properties of the plasma, but this would involve enormous amounts of computing time. The approach usually adopted then is somewhere between these two approaches, *i.e.* a two-dimensional cylindrical hybrid model. The more predictable parts are treated with general equations and the more intractable parts, those far from equilibrium, particularly excitation, are treated with Monte Carlo simulations using Newton's laws and random numbers. Each part needs the results of the others to begin, so the hybrid model must be solved iteratively, each part giving better approximations to the other parts until the whole converges.[7-9]

In the models, the argon gas atoms are assumed to be at rest and uniformly distributed throughout the cell with a number density determined by the pressure and gas temperature, using the ideal gas law. Fast electrons (*i.e.* electrons emitted from the electrodes) are treated with Monte Carlo simulation; slow electrons (bulk electrons, less than a few eV, too low for ionisation or excitation) and ions are treated with fluid equations. In the CDS, electrons, argon atoms and ions, and sputtering atoms, are treated with Monte Carlo methods. Outside the CDS, atoms and ions are dealt with by fluid equations.

Typical collisions included in the Monte Carlo models are elastic collisions, electron impact ionisation and excitation, fast argon impact ionisation and excitation, symmetric and asymmetric charge transfer, Penning ionisation (collisions with metastable argon atoms), ion–electron recombination, and de-excitation collisions.[4] Sputter yields come from semiempirical sputtering models such as by Matsunami *et al.*[10]

The models provide information on particle density profiles, particle energy distributions electric field distributions, the electric current (for a given voltage and pressure), the relative importance of the different collision processes, sputtering rates, level populations and intensities.[11] Much of this information is not available experimentally. How well the models agree with the available experimental data depends, of course, on the sophistication of the particular model used, but they all very much also depend on the values of a large number of input parameters, many of which are known only approximately. Some of the input parameters are pressure, gas temperature, various collision cross-sections, ionisation energies, atomic energy levels and transition probabilities, secondary electron yields, gas diffusion coefficients and sticking coefficients for atoms striking the sample and walls of the cell.[12,13] The gas

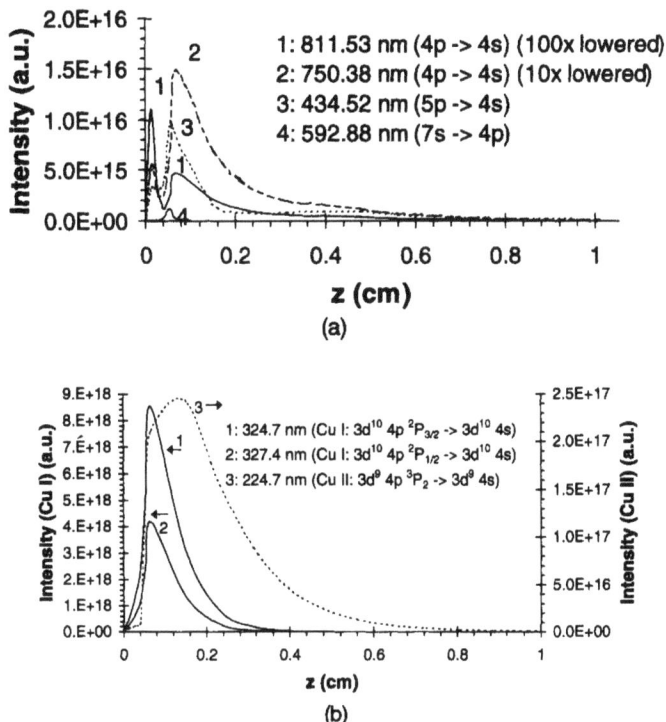

1: 811.53 nm (4p -> 4s) (100x lowered)
2: 750.38 nm (4p -> 4s) (10x lowered)
3: 434.52 nm (5p -> 4s)
4: 592.88 nm (7s -> 4p)

(a)

1: 324.7 nm (Cu I: $3d^{10} 4p\ ^2P_{3/2} \to 3d^{10}\ 4s$)
2: 327.4 nm (Cu I: $3d^{10} 4p\ ^2P_{1/2} \to 3d^{10}\ 4s$)
3: 224.7 nm (Cu II: $3d^9 4p\ ^3P_2 \to 3d^9\ 4s$)

(b)

Figure 11.4 *Calculated spectral line intensities as a function of distance from the cathode: (a) argon lines, (b) copper lines*[11]
(Reproduced with permission from A. Bogaerts and R. Gijbels, *J. Anal. Atom. Spectrom.*, 1998, **13**, 721)

temperature in the cell, for example, is not well known, is not constant throughout the cell and probably varies with changing plasma conditions. Along with the pressure, the gas temperature determines the gas density in the cell, a parameter critical to all the models.

For a copper sample on a Grimm-type source running at typical DC conditions, 800 V, 28 mA, 500 Pa, the models show that Ar and Cu intensities peak at a distance of about 1 mm from the sample surface, *i.e.* at the beginning of the NG[11] (see Figure 11.4). The intensities drop to near zero about 10 mm from the sample. Some Ar lines also have a peak in intensity very close to the sample, resulting from fast argon ion collisions. Most of the strong Ar emission appears in the range 700–1000 nm, *i.e.* red and near-IR, while most of the Cu lines were in the range 200–400 nm, *i.e.* UV. There is also a series of Ar blue lines (394.9–434.5 nm) responsible for the characteristic blue–violet colour of the glow discharge.[14]

The length of the CDS decreases markedly (from about 0.9 mm to about 0.55 mm) as the pressure is increased from 300 Pa to 700 Pa but decreases more slowly with increasing voltage. Typical particle densities obtained from the models are shown in Table 11.1.

Table 11.1 *Peak densities calculated for various species in an argon plasma at conditions typical in GDOES*[15]

Species	Peak density (m^{-3})	Relative to Ar atoms
Ar atoms	1×10^{23}	1
Ar ions	1×10^{20}	1×10^{-3}
Slow electrons	2×10^{20}	5×10^{-2}
Fast electrons	2×10^{15}	5×10^{-7}
Cu atoms	7×10^{20}	2×10^{-2}
Cu ions	3×10^{19}	3×10^{-3}

The mean energy of the electrons or ions is low but there is a high-energy tail in their distribution functions. The mean energies of the fast electrons in the tail are typically about 70% of the applied voltage, and are maximum at the end of the CDS, where they increase rapidly with increasing voltage but increase slowly with increasing pressure. The mean energies of the argon ions in the tail are maximum at the cathode surface (typically 100–300 eV, or about 30% of the discharge voltage) and increase rapidly with voltage but slowly with pressure. The mean energies of the bombarding argon atoms (typically 20–70 eV) are also maximum at the cathode surface and increase rapidly with voltage but slowly with pressure. Most of the argon ionisation is by fast electrons in the NG, having sufficient energy. The ions then enter the CDS with nearly thermal energies, gathering energy in traversing the CDS.

It might be expected that the mean energy of argon ions would decrease with increasing pressure, because of the increasing number of collisions they suffer. This effect is more than compensated for by the decrease in thickness of the CDS, finally leading to fewer collisions and slowly increasing mean energy.

Sputtering is caused by argon ions, fast argon atoms and by returning fast sample ions.[15] About 60–70% of sputtering atoms are redeposited onto the sample surface, so that net sputtering rates are only about 30–40% of the total sputtering.

The most important excitation mechanism is electron impact excitation.[16] Most excited states of sputtered atoms de-populate (de-excite) by radiative decay, *i.e.* by emitting photons.[17] But the excited states are not in local thermodynamic equilibrium (LTE). The complex balance in the plasma of collisional excitations (including metastable argon) and radiative decay means that the level populations of excited states do not match closely those predicted by the Boltzmann or Saha–Eggert equations, though, as expected, populations do tend to decrease exponentially with excitation energy. In particular, the relative populations of some atomic and ionic excited states may be quite different from those predicted from an equilibrium model.[17]

RF Plasma

To model an RF plasma, the calculations are repeated at each point in the RF cycle. In this way all the processes are monitored during the changing electric field caused by the RF voltage. Because of the DC bias developed on the surface of the sample, the

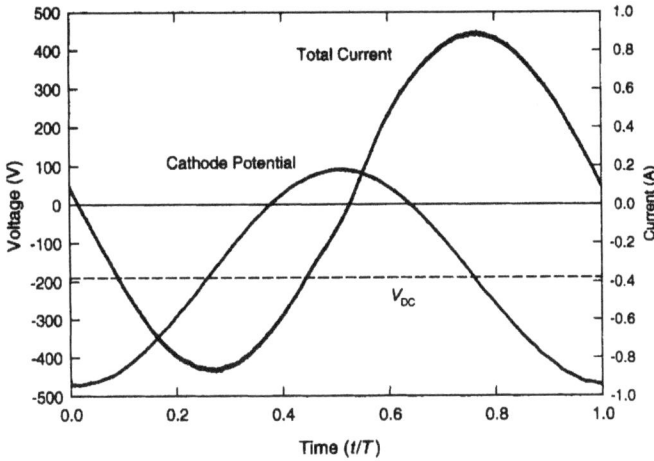

Figure 11.5 *Measured RF voltage and total current waveforms in an RF source (data from Belenguer et al., for a Ti sample at 5 W and 950 Pa[19])*

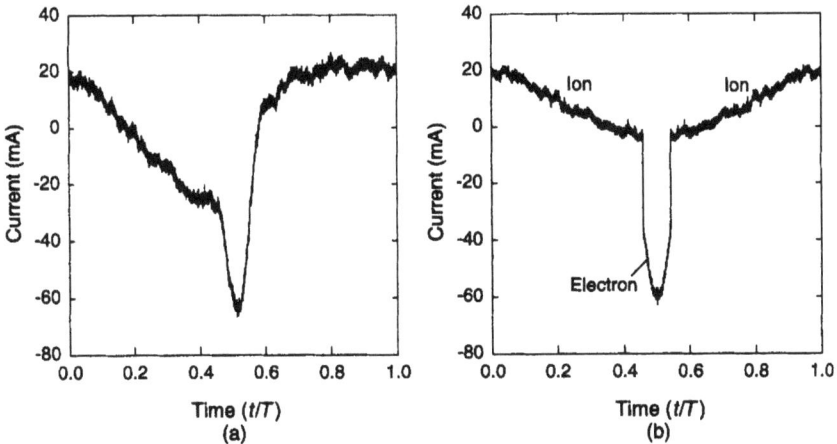

Figure 11.6 *Measured plasma currents in an RF plasma: (a) plasma current, (b) ion and electron currents (data from Belenguer et al., for a Ti sample at 5 W and 950 Pa[19])*

cathode voltage is positive only for a small part of the cycle (see Figure 11.5) where the positive excursion is shown in the middle of the cycle. From the capacitive nature of the cell, the voltage and total current (measured external to the source) are nearly 90° out of phase with each other. Most of the total current is displacement current, *i.e.* due to capacitive coupling and not due to charged particles flowing in the plasma.[18]

Once this displacement current is removed, it is then possible to see a more complex, much smaller plasma current; see Figure 11.6a. The plasma current now appears non-symmetrical due to another displacement current, this time caused by the temporal variation of the electric field inside the cell.[20] Once this additional displacement current is removed digitally, we have the ion and electron currents in the plasma; see Figure 11.6b. During most of the cycle, the current is dominated by ion current and only for a short time, during the positive part of the voltage cycle, by a strong electron current. This behaviour of the currents reflects the resistive nature of the analytical GD plasma.[21]

The power dissipated in the plasma is given by[22]:

$$P = \frac{1}{T} \int_0^T U_{pl}(t) I_{pl}(t) \, dt \qquad (11.1)$$

where T is one period. The plasma voltage U_{pl} is sinusoidal and given by:

$$U_{pl} = U_{HF} \sin(\omega_{HF} t) + V_{DC} \qquad (11.2)$$

where U_{HF} is the applied voltage, ω_{HF} is the angular frequency, t is time, V_{DC} is the DC bias voltage while the plasma current I_{pl} is the sum of its three components:

$$I_{pl} = J_{ion,rf} - J_{elec,rf} + J_{d,rf} \qquad (11.3)$$

Averaged over a full cycle, the total current is zero. This is why an RF plasma is possible with both conductive and non-conductive samples, since, unlike the DC case, there is no net current that has to flow through the sample.

During each cycle the CDS expands and contracts. At 13.56 MHz, one cycle takes 74 ns. A fast electron with an energy of perhaps 15 eV travels at 2300 km s^{-1} and is capable of traversing 85 mm in half a cycle, a distance far greater than the size of the cell. A fast argon ion, on the other hand, with perhaps 100 eV travels at 22 km s^{-1} and will traverse only 0.8 mm in half a cycle, maybe just enough time to traverse the CDS. The ions are therefore perturbed by the changing RF field, but their inertia means that they also tend to follow the average field, *i.e.* the DC bias voltage. The velocity of argon ions hitting the sample surface therefore varies during each cycle (see Figure 11.7). Thermalised ions travelling in the negative glow at perhaps 500 m s^{-1} will travel only 18 μm in half a cycle and so will be progressively slowed and speeded up during each cycle.

Because of the fluctuating electric field, slow thermal electrons can be heated up again (so-called wave-riding electrons, not important in GDOES), giving rise to so-called α-ionisation collisions.[24] This does not happen in DC. An RF plasma is more efficient at ionisation than a DC plasma and so can operate at lower voltages. For the same power and pressure, the voltage is lower in RF than in DC and the current is higher; this means the CDS in RF is thinner than in DC. Despite these differences many of the plasma properties are similar in RF and DC, resulting in similar intensities.[23,25]

Hydrogen Effect

Hydrogen has a large effect on our argon plasma. It enhances intensities from some emission lines, *e.g.* O 131 and N 149.26, and reduces intensities from others, *e.g.* Fe 371.99 and Cr 425.43; it increases spectral background and affects sputtering rates and crater shapes.[26–28] The high rates of conversion, both ways, between H atoms and H_2 molecules in the plasma mean that it does not seem to matter whether the

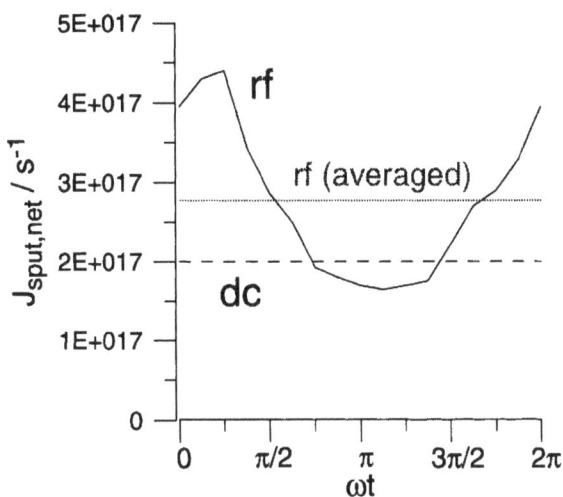

Figure 11.7 *Calculated net sputtering flux during an RF cycle (solid line)*[23]
(Reproduced with permission from A. Bogaerts and R. Gijbels, *J. Anal. Atom. Spectrom.*, 2000, **15**, 1191)

hydrogen comes from contamination of the argon gas or from sputtering from the sample. Possible hydrogen species in the plasma include ions, ArH^+, H_2^+, H_3^+, and neutrals, H, and H_2. The models cannot yet fully explain all the observed changes in intensities, but they do suggest that sputtering should decrease with increasing hydrogen, that ionisation of sputtered atoms should decrease because of reduction in Penning ionisation, the electron density should decrease because of recombination with hydride ions and the density of metastable argon should decrease because of quenching by H_2.[29]

Secondary Electron Yield

A plasma is generally understood as a mixture of neutral atoms or molecules and positively and negatively charged particles, usually positive ions and electrons. The overall charge of the plasma is zero, the number of positively and negatively charged particles being equal. The particles of opposite charge will spontaneously recombine to form neutral species, so processes are needed to create new charged particles to maintain the plasma. The emission of secondary electrons, *i.e.* electrons leaving the cathode as a consequence of the ion and neutral bombardment of the sample surface, is one such crucial process.

The secondary electron yield, γ, is the average number of electrons emitted per incident ion. Empirically it has been found that γ increases as the work function ϕ decreases, *i.e.*:[30–35]

$$\gamma = 0.032\,(0.78E_i - 2\phi) \qquad (11.4)$$

where E_i is the effective potential energy (for argon bombardment, assumed to be the argon ionisation energy, 15.76 eV). The work function of a material can vary with

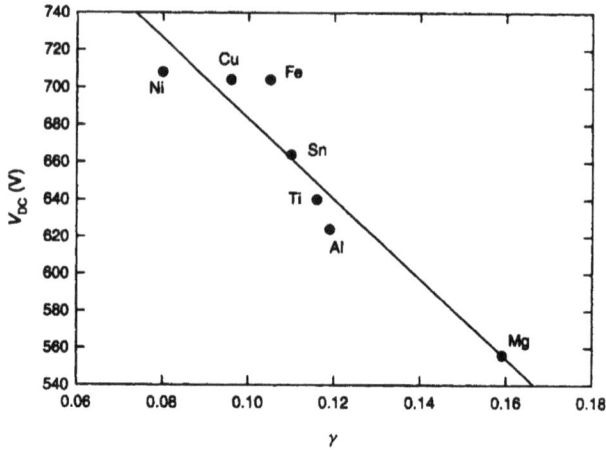

Figure 11.8 *DC bias voltage measured on nearly pure metals as a function of their calculated secondary electron yield, γ*

the matrix composition, crystal structure and surface preparation. Estimates of work functions vary significantly, especially for elements with low ionisation potential, and care is required when the term in parentheses is very small.[36] Hence, the estimates for γ given by Equation 11.4 are indicative of general trends only. When argon ions are used for bombarding the surface, γ is typically 0.1 for metals, but is estimated to vary from 0.054 for Pt to 0.16 for Mg,[33,37] *i.e.* by about a factor of 3.

A secondary electron emission yield of 0.1 means that we need ten ions to create one new electron; in fact we lose nine ions in the process and on an average as many electrons to the anode. To counterbalance this loss in charged particles, the electron liberated at the sample surface must regenerate the lost ion–electron pairs, mainly through collisions. We can therefore easily understand that the secondary electron emission yield is directly linked to the power needed to maintain a given charge density.

The impedance of the GD source will change with the number of electrons, *i.e.* with γ. At constant power and pressure, a higher value for γ will mean a lower voltage and a higher current. To verify this, the DC bias voltage was measured in RF operation, at constant applied power (50 W) and pressure (600 Pa), using the purest materials available. The results in Figure 11.8 show the expected trend: DC bias voltage decreases with increasing secondary electron yield. But, to complicate matters, it is also possible that a higher γ could lead to a lower gas temperature, producing a higher gas density, equivalent to a higher pressure, and hence a lower DC bias.

Source Impedance

The impedance of an analytical GD plasma is determined by the electron emission characteristics of the cathode (sample), the argon pressure in the source and the cell

Table 11.2 *Source impedance for different materials, at constant pressure*

Sample	Incident power (W)	Cell resistance (Ω)	Cell capacitance (pF)
Steel NBS 1762	35	4760	32
Steel NBS 1762	50	4710	32
Al-Si Pechiney 6013	35	4280	34
Ceramic CC650A	35	4930	30

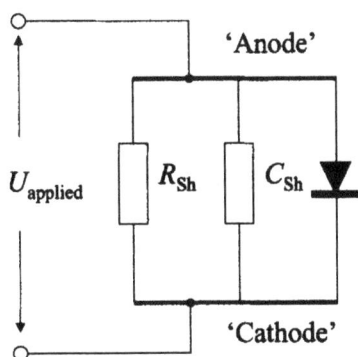

Figure 11.9 *Simplified equivalent circuit for an analytical GD plasma*

geometry (including the anode-to-cathode distance). Because of the large ratio in the surface areas of anode to cathode, the resulting plasma is dominated by the cathode sheath. In its simplest form the analytical GD plasma can therefore be represented by the equivalent circuit shown in Figure 11.9. R_{Sh} is typically some thousands of ohms and C_{Sh} is about 0.2 pF.[38]

The coaxial connection, from the RF power source, and the cell geometry of the source create a stray capacitance C_{stray} typically some tens of picofarads. This acts in parallel with the plasma impedance so that the source capacitance is the sum of the plasma capacitance and the stray capacitance and hence is dominated by the stray capacitance. Non-conductive samples will add an additional series capacitance from the dimensions and dielectric properties of the sample.

For fixed argon pressure and cell geometry, the source resistance R_{Sh} will change with changes in the cathode material. Table 11.2 shows measured values for some common materials mounted on a GD source, at constant pressure. There is no significant change in R_{Sh} for the steel sample when the power is increased from 35 W to 50 W. The small change in capacitance is related to the change in sample diameter; the smallest sample gives the smallest cell capacitance.

For fixed geometry and the same sample, the impedance will decrease with increasing pressure. For fixed argon pressure and the same sample, the impedance will decrease with increasing anode-to-sample gap.

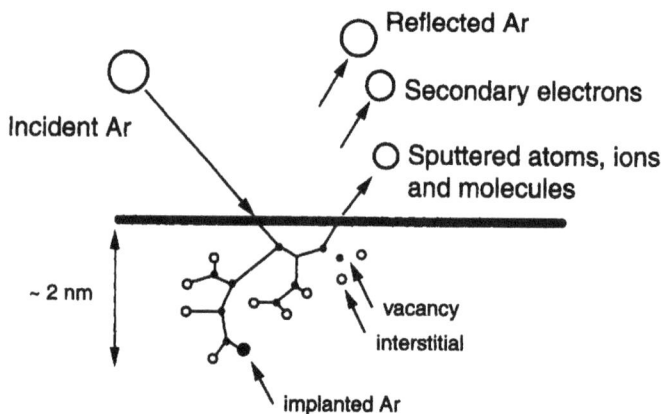

Figure 11.10 *Illustration of events accompanying sputtering of a surface by argon bombardment*[42]
(Reproduced with permission from B.V. King, *Glow Discharge Optical Emission Spectrometry,* John Wiley & Sons, Chichester, 1997, 245–53)

3 Sputtering

Sputtering is the ejection of sample atoms following the impact of energetic ions (and neutral atoms) on the surface. An incoming ion suffers a binary collision with a target atom. From simple mechanics, the maximum fraction η of the incoming energy that can be transferred to the target atom during a head-on binary collision is given by[39]:

$$\eta = \frac{(m_i + m_t)^2}{4m_i m_t} \qquad (11.5)$$

where m_i and m_t are the masses of the incident and target particles respectively. The function η increases rapidly as m_t approaches m_i, is a maximum at $m_t = m_i$ and then declines slowly at higher values of m_t. For incident argon, this means that the optimum energy transfer occurs for elements with atomic numbers near 40. Following the impact, the energy is spread throughout the neighbouring atoms, one or more of which receives enough energy to break its bonds and escape (see Figure 11.10). This spread of energy means that, on average, only about 1/16 of the transferred energy is gained by the sputtered atom. This means that the transferred energy must be about 16 times the average bond energy before sputtering can occur. The heat of sublimation, U_S, is a measure of the average energy required to break the atomic bonds in the sample. U_S is typically 3–9 eV.[40] We can therefore estimate an average threshold energy for sputtering, E_0, given by:

$$E_0 = 16U_S/\eta \qquad (11.6)$$

For incident argon atoms and ions, with $m_i = 40$ amu, E_0 varies from 31 eV for lead to 152 eV for aluminium, and is typically around 100 eV. In the Grimm source, the minimum potential for sputtering, U_0, in DC, typically varies over the range of 250–380 V, depending on the sample matrix, and has a mean of about 308 V.[41] It is

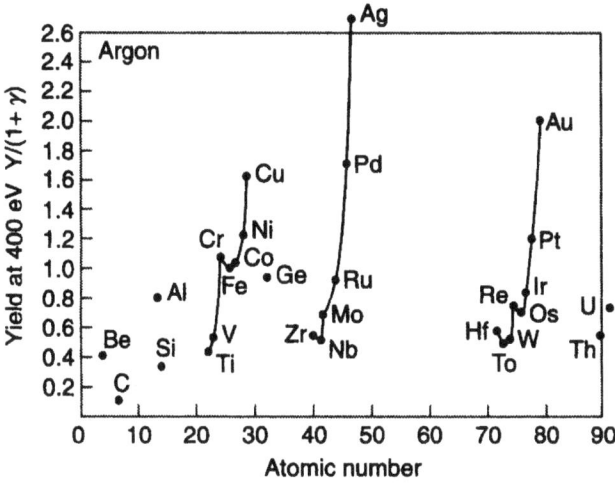

Figure 11.11 *Sputter yields of various elements for an Ar bombardment of 400 eV*[43] (Reproduced with permission from A. Benninghoven *et al.*, *Secondary Ion Mass Spectrometry*, John Wiley & Sons, New York, 1987)

much higher than the energies estimated from Equation 11.6 because the incoming argon ions lose much of their energy through collisions in the plasma before striking the sample surface. In RF the minimum potential is lower, about 150 V.

The sputtering yield, S, is the number of sputtered atoms per incident ion. The net sputtering rate, q_M, is related to S by:

$$q_M = d\frac{Si_g}{(1+\gamma)} \tag{11.7}$$

where d is the fraction of sputtered atoms not redeposited (d is not well known but may typically be about 0.4, varying with the matrix and the argon pressure), and γ is the secondary electron yield (typically about 0.1). The sputtering yield S varies with the impact energy (see Figure 11.11). Multiple scattering and anisotropic scattering make the sputtering yield non-linear with impact energy.

Although it can be expected that sputtering rates in a GDOES will depend on basically any parameter influencing the plasma, such as voltage, current, pressure, trace gases and the cathode material itself, it has been found empirically that sputtering can be well approximated by:

$$q_M = C_Q i_g (U_g - U_0)^N \tag{11.8}$$

where C_Q is a sputtering constant which may vary with the plasma gas species but not with current, potential or pressure, U_0 is a constant and $N = 0.74$. Equation 11.8 is a non-linear version of the equation first used in GDOES by Boumans.[44] Note that $N = 1$ in Boumans' original equation, which is still used when sputtering rates are measured with constant time. Note, also, that in calculating relative sputtering rates, discussed in more detail in the next section, the off-set voltage U_0 is usually considered to be independent of the sample material (typically 300 V in DC when $N = 1$, and lower in RF), which, however, is not quite true.

At constant pressure, the sputtering rate can also be expressed as:

$$q_M = C_Q(P_g - P_0) \tag{11.9}$$

where P_g is the applied power and P_0 is a constant that may vary with the sample matrix. Since P_0 is generally quite small, perhaps 2–3 W in RF, the sputtering rate is nearly proportional to power. At higher applied RF powers of 30–50 W, P_0 can often be ignored.

When several elements are present in the target, the surface will become preferentially enriched with the slower sputtering elements, and these will dominate the sputtering rate constant. The sputtering rate for a material composed of different elements is therefore given approximately by[45]:

$$\frac{1}{q_M} = \frac{1}{100} \sum_i \left(\frac{c_i}{C_{Q_i}} \right) \tag{11.10}$$

where c_i is the content of element i (in mass %).

An important parameter in GDOES is the elemental sputtering rate q_i. This varies with the composition, c_i, of element i in the sample and with the overall sputtering rate, q_M, and is given by:

$$q_i \equiv c_i q_M \tag{11.11}$$

Sputtering rates, *i.e.* the masses removed per second, in GDOES do not appear to vary significantly with anode diameter. The sputtered depth, on the other hand, will change with anode diameter because approximately the same mass s^{-1} is being removed from different volumes of the sample. For a constant mass removal rate, the depth should vary as the inverse square of the inner diameter. A more interesting way to present sputtering rates, therefore, rather than mass per second, is mass per unit area per second. In this way sputtering equations do not depend explicitly on the anode diameter.

Relative Sputtering Rates

Relative sputtering rates appear not to depend strongly on changing plasma conditions, at least within experimental error and over the normal range of conditions used for analysis. At constant current and voltage, from Equation 11.8:

$$\frac{q_M}{q_{Ref}} = \frac{C_{QM}(U_g - U_{0M})^N}{C_{QRef}(U_g - U_{0Ref})^N} \approx \frac{C_{QM}}{C_{QRef}} \tag{11.12}$$

suggesting relative sputtering rates should be constants that are insensitive to changes in plasma conditions, provided the applied voltage is not too close to U_0. At constant applied power and pressure, from Equation 11.9:

$$\frac{q_M}{q_{Ref}} = \frac{C_{QM}(P_g - P_{0M})}{C_{QRef}(P_g - P_{0Ref})} \approx \frac{C_{QM}}{C_{QRef}} \tag{11.13}$$

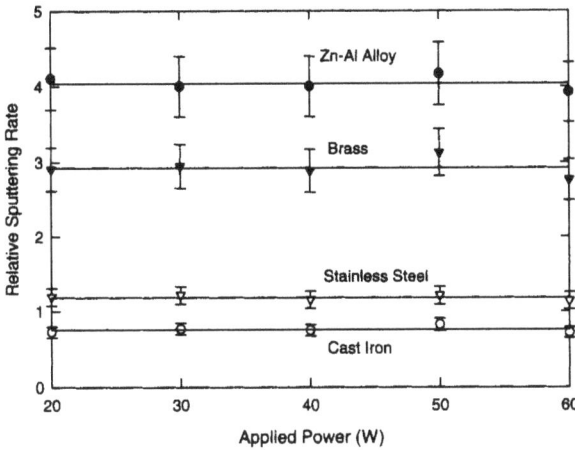

Figure 11.12 *Measured sputtering rates relative to pure iron, for five different metals, at a constant argon pressure of 700 Pa but with variable applied power. (Bars shown are for 10% uncertainty)*

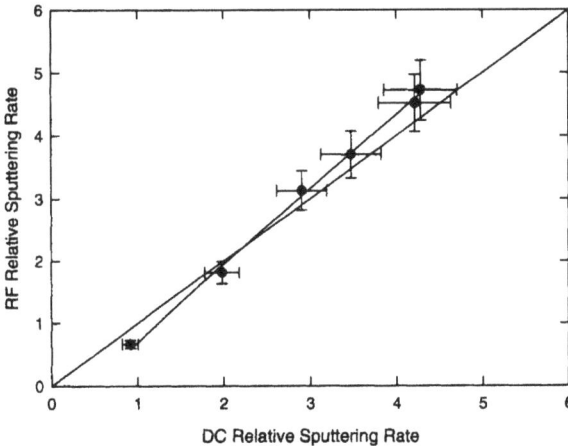

Figure 11.13 *Relative sputtering rates for six certified Zn–Al alloys, measured with RF and DC.*

again suggesting that relative sputtering rates are nearly constant. In Figure 11.12, relative sputtering rates are shown measured at fixed pressure (700 Pa) over a range of applied RF powers; similar results are found if the sputtering rates are plotted with fixed applied power and variable pressure.

Relative sputtering rates measured with RF appear to be similar to those measured with DC, see Figure 11.13. The reason for the apparent trend in the figure, between RF and DC, is not yet clear and may be due more to differences between laboratories in measuring sputtering rates with systematic trends in crater shape rather than to

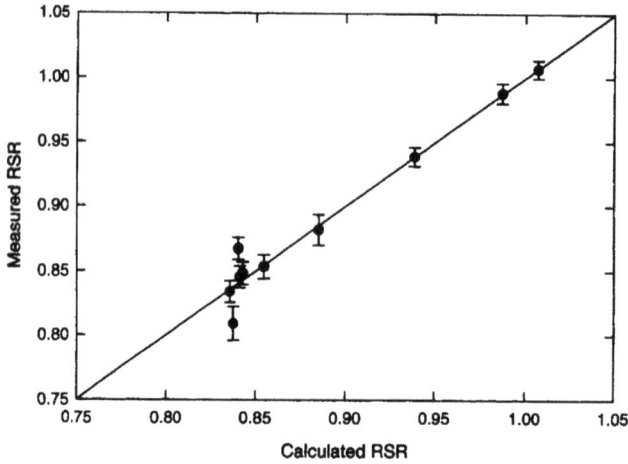

Figure 11.14 *Comparison of calculated and measured relative sputtering rates for white cast iron*

real differences in the sputtering rates between RF and DC, or may be due to the assumptions neglecting changes in U_0 and P_0 inherent in Equations 11.12 and 11.13.

When Equation 11.10 is combined with semiempirical ion-scattering theory, the sputtering rate $q_{M\mathrm{rel}}$ of a material, expressed as a ratio to some reference material, assumed to be a single element, *e.g.* pure iron, is given approximately by[45,46]:

$$\frac{1}{q_{M\mathrm{rel}}} = k''_{12} \sum_i (c_i \eta_i U_{\mathrm{Si}1}) \tag{11.14}$$

where k''_{12} is a constant, $\eta_i = \frac{(m_i + m_{\mathrm{Ar}})^2}{m_i^{2.4}}$, where m_i, m_j and m_{Ar} are the atomic masses of the element i in the material, the element j in the reference material and argon (the sputtering gas), $U_{\mathrm{Si}1}$ is the sublimation energy of the element i in the material. In other words, the relative sputtering rate should vary with the mix of atomic masses in the sample and with their sublimation energies. Equation 11.14 therefore represents another way forward for determining the relative sputtering rates of unknown samples.

It is not yet possible to use this equation directly for all materials because $U_{\mathrm{Si}1}$ is a property of 'an element in a matrix' and will vary in an as-yet unknown way from one matrix to another. But if we restrict the equation to a single matrix, then $U_{\mathrm{Si}1}$ will be constant and the equation simplifies to:

$$\frac{1}{q_{M\mathrm{rel}}} = \sum_i (g_i c_i) \tag{11.15}$$

where $g_i = k''_i \eta_i U_{\mathrm{Si}1}$. That is, the sputtering rate is now a linear combination of the elemental composition, with g_i being constants that can be determined by regression to measured sputtering rates. Figure 11.14 shows results for white cast iron.

The case of materials varying in composition over large ranges, such as the wide range of zinc and aluminium alloys, is a more complex problem. In such cases, if the sublimation energy varies in a systematic way with composition, then Equation 11.15 can be modified to include this variation.[46]

While it is too soon to be sure how valuable it will be, a library of relative sputtering rates is being developed within the GDS community that, it is hoped, will work for all common plasma conditions without the need to remeasure them each time conditions are changed. The relative sputtering rates for some pure materials and some common alloys were given in Chapter 6.

Unlike relative values, absolute sputtering rates will change with changing plasma conditions, but this change in absolute sputtering rate is accounted for in the sputtering rate of the reference sample. Hence, each time a new analytical method is created with different plasma conditions, the absolute sputtering rate of the reference sample should be reestimated for the conditions of the new method. This could be done by calculating the new reference sputtering rate from one of the sputtering rate Equations 11.8 or 11.9, or by remeasuring the sputtering rate of the reference sample with the new method.

4 Quantitative Analysis

Sputtered atoms leave the surface of the sample, diffuse through the plasma into the NG region where they are excited through energetic collisions and de-excite through emission of characteristic lines. The emission is then detected and recorded by the detectors in the spectrometers attached to the glow discharge source.

There are therefore three processes in generating the analytical signal:

1. the supply of sputtered atoms
2. excitation followed by photon emission
3. detection

To understand the procedure of quantitative analysis it helps to keep in mind that the detectors look into the plasma and not into the sample. Sputtering of material from the sample to the plasma allows us to make a link between the plasma and the sample.

To simplify things it is normally assumed that these processes of sputtering and emission are independent. Hence, the recorded signal for a given emission line from element i is given by:

$$I_i = k_i e_i q_i \qquad (11.16)$$

where q_i is the supply rate of element i into the plasma, e_i represents the emission process and k_i is the instrumental detection efficiency. The supply rate is equal to the elemental sputtering rate.

The emission term will vary with the number of photons emitted per sputtered atom and with the absorption of these photons in traversing the plasma to reach the source window, so that:

$$e_i = S_i R_i \tag{11.17}$$

where R_i is the emission yield (defined here as the number of photons emitted per sputtered atom entering the plasma) and S_i is the correction for self-absorption, discussed below, that varies between 0 and 1 depending on the nature of the emission line and the elemental sputtering rate. The detection efficiency, which includes the angle subtended by the source window, the transmission through the window, the grating efficiency, the photomultiplier or solid-state detector, and the detection electronics, is assumed constant, as variations in spectrometer response are not considered here. To these must be added a background term, b_i, originating from detector dark current, instrument noise, scattered light, argon emission and unwanted signals from nearby emission lines. These considerations lead to the following general equation for GDOES[47,48]:

$$I_i = k_i S_i R_i c_i q_M + b_i \tag{11.18}$$

Equation 11.18 in fact represents a set of equations, one for each element i in the sample.

For analytical convenience, and introducing relative sputtering rates, the equation can be rewritten as:

$$c_i q_M / q_{Ref} = k'_i S'_i R'_i I_i - b'_i \tag{11.19}$$

where $'$ means that the terms are inverted, *e.g.* R'_i is now the inverse emission yield. The term $c_i q_M / q_{Ref}$ is now the relative number of atoms per second of a particular element entering the plasma.

5 Compositional Depth Profiling

In any analysis or at any particular depth in a depth profile, when the signals from all the elements with significant compositions are recorded, it can be assumed the that compositions will add up to 100%, *i.e.*:

$$\sum_i c_i = 1 \tag{11.20}$$

where c_i is in units of atomic or mass fraction.

During a depth profile we measure intensities as a function of time (see Figure 11.15a), where the data are from Ishibashi and Yoshioka.[49] From calibration equation (11.19) we know the relationship between intensity and $c_i q_M / q_{Ref}$, *i.e.*, at some time t_1:

$$\{c_i q_M / q_{Ref}\}_{t_1} = \{f(I_i)\}_{t_1} \tag{11.21}$$

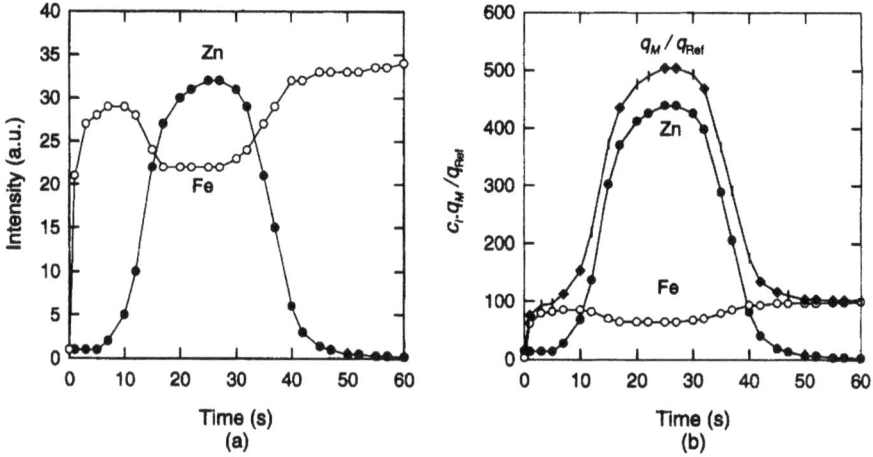

Figure 11.15 *Initial process from qualitative to quantitative: (a) qualitative, (b) generate* $c_i q_M/q_{Ref}$ *and* q_M/q_{Ref}

where $f(I_i)$ is the calibration function for element i. Hence at time t_1 during the depth profile we know the relative sputtering rate since:

$$\{q_M/q_{Ref}\}_{t_1} = \left\{\left(\sum c_i\right)(q_M/q_{Ref})\right\}_{t_1} = \left\{\sum (c_i q_M/q_{Ref})\right\}_{t_1} = \left\{\sum [f(I_i)]\right\}_{t_1}$$

(11.22)

See Figure 11.15b.

Now we can calculate all the compositions at time t_1 from

$$\{c_i\}_{t_1} = \{f(I_i)\}_{t_1} / \{(q_M/q_{Ref})\}_{t_1}$$

(11.23)

In fact we have simply normalised the sum of all $c_i q$ to 1.

We can calculate the sputtered mass during the time interval Δt at time t_1 from:

$$m_{t_1} = q_{Ref} \{(q_M/q_{Ref})\}_{t_1} \Delta t$$

(11.24)

To find the total mass loss M_{t_1} from time 0 to time t_1, we simply sum the values of m_t from m_0 to m_{t_1}:

$$M_{t_1} = \sum_{t=0}^{t_1} (m_t) = q_{Ref} \sum_{t=0}^{t_1} [\{(q_M/q_{Ref})\}_t \Delta t]$$

(11.25)

In Figure 11.16a we have plotted composition as a function of mass loss. Note that unlike depth, which follows, mass loss does not contain any assumption about density.

To determine the depth traversed during the time interval Δt at time t_1:

$$\Delta z_{t_1} = \frac{m_{t_1}}{\rho_{t_1}}$$

(11.26)

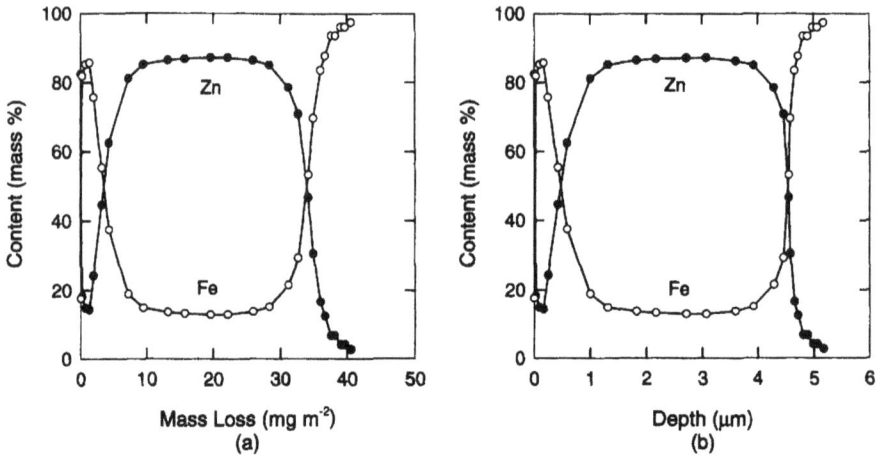

Figure 11.16 *Final process from qualitative to quantitative: (a) content and mass loss, (b) depth*

To find the total depth from time 0 to time t_1, we simply sum the values of Δz_t from Δz_0 to Δz_{t_1}:

$$z = \sum_{t=0}^{t_1} (\Delta z_t) = q_{Ref} \sum_{t=0}^{t_1} \left[\{(q_M/q_{Ref})\}_t \frac{\Delta t}{\rho_t} \right] \qquad (11.27)$$

The resulting CDP is shown in Figure 11.16b.

The term q_{Ref} should be determined by measurement but examination of Equation 11.27 suggests that it could act like an adjustable parameter for the total depth, in which case the estimation of uncertainty in depth is very much a question of personal choice.

So, once the calibration is done, the whole process is a simple one. At each moment during the depth profile, we convert the measured intensities I_i into their $c_i q_M/q_{Ref}$ values from the calibration, then sum the $c_i q_M/q_{Ref}$ to get the relative sputtering rate q_M/q_{Ref}, then divide the individual $c_i q_M/q_{Ref}$ values to get the compositions c_i and integrate the q_M/q_{Ref} values from time zero, correcting for density, to get the depth.

Emission Yield

The most troublesome term in Equations 11.18 and 11.19 is the emission yield, R_i. The emission yield is the term that relates the number of photons coming from the source to the number of atoms entering the plasma. If it is constant, then counting the photons is equivalent to counting the number of atoms in the plasma. It is, therefore, only changes in emission yield that upset this counting process.

The excitation of sputtered atoms in the NG region can occur through collisions with relatively high-energy electrons or through collisions with metastable argon atoms.[3] The emission yield should therefore increase with current since more current means more electrons for collisions and the generation of more metastable argon atoms. The emission yield might be expected to decrease with voltage since electron collision cross-sections decrease at higher energies. The variation of R_i with pressure is more complex. At low pressures, the number of metastable argon atoms increases with pressure because there are more argon atoms available to be excited into their metastable state; but above about 250 Pa (1 Torr), the number of metastable argon atoms starts to decrease because an increasing number suffer collisions with other argon atoms causing them to drop out of their metastable state.[50] Therefore, at typical analytical conditions, pressure >250 Pa, the emission yield for electronic states being excited by collisions with metastable argon atoms should decrease slowly with increasing pressure.

One way to express these general trends with changing plasma conditions is[51]:

$$R_i = i_g^a (U_g - U_0)^b p_g^c \qquad (11.28)$$

where U_0, a, b and c are matrix-dependent constants; a is typically about 1, b is about -0.5 and $c < 0$. The actual values of a, b and c vary from element to element, and from line to line within each element. Historically, this equation was established using the experimental values obtained with DC sources but there is no reason to suppose that an RF source would be essentially different.

An earlier and widely used approach is to assume that the pressure dependence is negligible and then avoid the necessity for estimating U_0 by expanding the voltage dependence as a third-order polynomial[52]:

$$R_i = i_g^a \left(b_0 + b_1 U_g + b_2 U_g^2 + b_3 U_g^3 \right) \qquad (11.29)$$

An interesting feature of emission yield often overlooked is that if the pressure is kept constant and the power varies, *e.g.* by varying the voltage, then the decrease in emission yield due to the increased voltage is very nearly balanced by the increase in emission yield due to the increased current. This means that it should be possible to do a calibration with fixed pressure, and variable power without the need to correct for emission yield changes.

During the operation of a GD source, there is always at least one parameter that is not fixed. In RF operation with constant power and pressure, it is the ratio of voltage to current; in DC operation with constant current and voltage, it is pressure. What largely determines the value of the free parameter for a particular sample is the plasma impedance (more precisely, the plasma resistance, R_g, which we saw earlier is largely the cathode sheath resistance). If we are discussing RMS values in RF operation or DC values in DC operation, we can think of impedance as the ratio of voltage to current. In RF operation at constant power, impedance will change with the DC bias voltage. At constant power, changes in the emission yield can therefore be expressed as some function of impedance, $R_i = f(R_g)$. Experience suggests that $R_i \sim R_g^d$, where d ~ -0.5.

It is helpful to have some feel for how much variation in emission yield could be expected for different operating conditions. What determines the impedance for a particular sample (all other things being constant) is the secondary electron yield. From Chapter 10, Section 3 we saw that the secondary electron yield, γ, can vary by about a factor of 3 for different metal elements. Because of this variation in electron yield, the current, should also vary by about a factor of 3, at constant voltage and pressure. Since emission yield is nearly proportional to current, the emission yield should also change by about a factor of 3.

For constant current and pressure, the voltage (or at least the effective voltage, *i.e.* the voltage above threshold) should vary by about a factor of 3. This can be seen in Figure 11.8, assuming a threshold voltage of about 300 V. Since the emission yield is nearly proportional to the square root of the effective voltage, the emission yield should then vary by about $\sqrt{3} \approx 1.7$.

For constant power and pressure, the voltage and current should each change by about 1.7, giving a combined change in emission yield of about 2.3. These are only estimates, and perhaps a worst case in the sense that we have assumed that γ could vary by a factor of 3 in our analysis. When we analyse a single matrix, in bulk analysis, for example, γ and hence emission yields would not be expected to vary significantly. But the variations described are what we might see when we combine many different matrices in a single calibration for CDP.

Given the normal ranges of these variables, the most important effect is varying current, next is varying voltage and thirdly varying pressure. If we keep current and voltage constant, then the variation in R_i because of varying pressure is at most only about 10–20%.

This work has been interpreted to mean that we should operate a glow discharge with constant current and voltage and variable pressure, to minimise the variation in R_i. This is possible in DC operation for metals and increasingly possible in RF for both metals and non-metals. But there are many technical advantages for keeping the pressure constant, especially at the surface and at interfaces where plasma conditions change quickly. Constant pressure has been the most successful mode to date for non-conductive coatings. For metallic coatings, similar results have been obtained for either constant or variable pressure, but with constant pressure it is necessary to correct for variations in impedance, *e.g.* by monitoring changes in DC bias voltage.

Self-absorption

Self-absorption arises from the absorption of photons by atoms positioned in the glow discharge between any emitting atoms and the exit window of the source. The effect is important to optical emission spectroscopy, as it tends to bend the calibration curves and make them non-linear. For self-absorption to occur, the absorbing atoms must be capable of absorbing at the wavelength of the emitted photon (see Figure 11.17). This generally means that they are of the same element as the emitter and they must be in the same energy level as the final state of the emission process.

To find a significant number of these atoms in a glow discharge plasma with its fairly low excitation density, the final state of the transition must be the ground state or very near to the ground state or in a long-lived metastable state. This explains why most spectral lines showing self-absorption are resonance or near-resonance lines. Resonance lines are defined as lines with a final energy level at the ground state, 0 eV; near-resonance lines are defined as lines with a final energy state above 0 eV but ≤ 0.25 eV (≤ 2000 wavenumbers).[54] Some transitions ending in long-lived metastable states may also show self-absorption.[55]

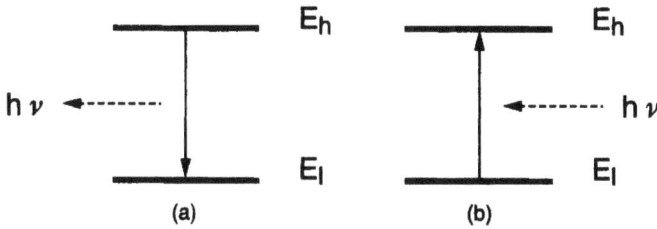

Figure 11.17 *Process of self-absorption*[53]
(Reproduced with permission from K. Wagatsuma, in *Glow Discharge Optical Emission Spectrometry,* R. Payling, D.G. Jones and A. Bengtson (eds), John Wiley & Sons, Chichester, 1997, 342)

As many resonance or near-resonance lines are analytically interesting, because of their good detection limits, it is not unlikely to run into these lines in GDOES. Moreover, if the absorbing atoms are in a cooler non-emitting region, as may be the case in the GD source since the electron density of the plasma decreases away from the sample towards the window, the absorption will be strongest at the centre of the emission line and self-reversal (*i.e.* loss of intensity at the centre of the line) may occur.

Optical Depth

The severity of self-absorption varies with a term called the optical depth. The greater the optical depth, the worse the self-absorption. This is why plasmas without significant self-absorption are described as 'optically thin'.

The optical depth, K_L, for a particular emission line from an element, is the product of an effective absorption cross-section s_L and the number of atoms or ions of the element in the plasma in the lower energy state of the optical transition. They have to be in this state ready to absorb a passing photon. The number of such atoms or ions is proportional to the rate at which they are supplied to the plasma. In GDOES this rate is proportional to the elemental sputtering rate, hence:

$$K_L \propto s_L \left(f_{il} c_i q_M \right) \tag{11.30}$$

where f_{il} is the fraction of the element i present in the lower energy state of the transition of interest. For typical GDOES conditions we can assume that $f_{il} \sim 1$ for the ground (or near-ground) state of atoms and that $f_{il} \sim 0$ for other levels, *i.e.* for ions and for the excited states of atoms, since the populations of these states in a GD plasma are generally much lower than the populations of the ground states of the atoms.

Historically, atomic lines are given by the symbol 'I'; the first ionic state (singly charged ions) by the symbol 'II' and higher ionisation states (multiply charged ions) by the symbols 'III', 'IV', 'V', *etc.* Most lines of interest in GDOES come from atomic or first ionic states, *i.e.* from 'I' or 'II' states. Transitions to the ground state are called resonance transitions, or resonance lines, and are given by the symbol 'R'.

Transitions to states near the ground state (≤ 0.25 eV) are called near-resonance lines, and given by the symbol 'r'. Hence, an atomic resonance line would be 'I R'.

The effective absorption cross-section is given by:

$$s_L = \sqrt{\frac{\pi}{2\ln(2)\,R}}\,\frac{10^4}{(4\pi)^2}\sqrt{\frac{M}{T}}\lambda_0^3\frac{g_j}{g_i}A_{ji} \approx 33.06\sqrt{\frac{M}{T}}\lambda_0^3\frac{g_j}{g_i}A_{ji} \qquad (11.31)$$

where M is the atomic weight, T is the absolute temperature of the plasma, λ_0 is the wavelength (in metres) at the centre of the line, g is the statistical weight of the level and A_{ij} is the atomic transition probability (in s^{-1}). The larger the s_L, the worse the self-absorption. This final equation allows us to use the extensive database of atomic transition probabilities created by Payling and Larkins.[54] Values for s_L for the most commonly used lines in GDOES are shown in Appendix A. Note that for practical considerations in GDOES, high values of s_L are only of concern for 'I R' and 'I r' states.

The statistical weight is $2J + 1$ where J is the total angular momentum quantum number of the level. For example, consider the resonance emission line Zn 213.857 $^1S_0-^1P°_1$. By convention the lower state is shown first, and J is the subscript. The value of J for the lower level is 0 and for the upper level it is 1. Hence, $g_i = 2 \times 0 + 1 = 1$ and $g_j = 2 \times 1 + 1 = 3$.

Exponential Model

To understand how self-absorption leads to non-linear calibration curves, let us look first at a very simple model of the plasma. We will consider a single homogeneous plasma, emitting photons and re-absorbing some of them (alternatively, one could imagine an optically thin emitting region followed by an absorbing region).

The emitted signal, without self-absorption, is:

$$I_{i0} = a_i c_i q + b_i \qquad (11.32)$$

The lost intensity on a very small path is:

$$dI = I\,s_s\,dz \qquad (11.33)$$

where s_s is a factor describing the optical density of the medium and z its thickness. For constant optical density, following the rule of Lambert–Beer on absorption of light, we can express this intensity passing through the plasma as:

$$I_i = I_{i0}e^{-s_s z} \qquad (11.34)$$

As the optical thickness increases with the number of absorbers, but the total thickness is constant, we can rewrite the equation as:

$$I_i = I_{i,0}e^{-s_e c_i q} \qquad (11.35)$$

Combining Equations 11.32 and 11.35, we find:

$$I_i = a_1 c_i q\, e^{-k_s c_i q} + b_i \qquad (11.36)$$

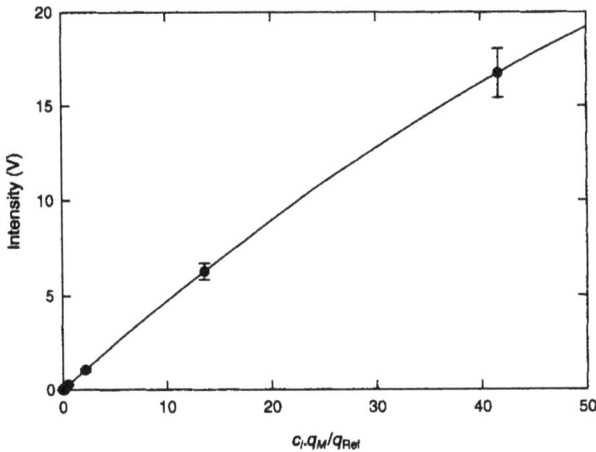

Figure 11.18 *Expected change in intensity of Ni 341 for increasing atom density $c_i q_M$ of element i, in the presence of self-absorption described by Equation 11.36*

where we have introduced a background term. The equation is illustrated in Figure 11.18, where the data for Ni 341 are from the calibration in Chapter 6. As the atomic density in the plasma increases, *i.e.* $c_i q$ increases, more and more self-absorption occurs and the intensity curve bends.

In reality the plasma is, of course, more complex than this simple model. Either the plasma is not homogeneous, or the emission and absorption regions will not be so nicely separated; but emitters and absorbers will both be present with varying relative and absolute densities. The emitters will decrease more rapidly than the absorbers as we move along the inside of the anode tube away from the peak in intensity in the NG.

We also need to take into account the nearly Gaussian profile of the spectral lines, because atoms with different velocity distributions will re-absorb photons preferentially matching their energy distributions. The kinetic temperature will not be constant over the entire plasma area. The particles, absorbers and emitters will certainly be cooler away from the sample. The Gaussian profile has a maximum at its centre, speed zero relative to the detector. This means that self-absorption in a cold region will be stronger in the centre of the Gaussian profile than in a hot region.

Model Including Gaussian Profile

Modelling the glow discharge as an inner emitting/absorbing region and an outer absorbing region, where the probability of self-absorption increases exponentially with the number of absorbers and the Gaussian profile of the spectral lines is taken into account, leads to the following approximation:

$$I_i = \frac{k_i c_i q_M}{1 + 0.412 s_E c_i q_M} \exp\left(-s_S c_i q_M\right) + b_i \qquad (11.37)$$

where I_i is the measured intensity, k_i is a calibration constant, c_i is the composition in the sample, q_M is the sputtering rate, b_i is the background signal, s_E is the self-absorption coefficient for the inner emission/absorption region and s_S is the coefficient for the outer absorbing region.[56,57] Typically, $s_S \sim 0.1s_E$, *i.e.*, most of the absorption occurs in the emission region where the density of absorbers is higher.

We can expand the right hand side as a polynomial, and invert the polynomial to be powers of I_i to obtain our analytical calibration function:

$$c_i q_M = K_i I_i + 0.4502 s_i K_i^2 I_i^2 + 0.2653 s_i^2 K_i^3 I_i^3 + 0.1515 s_i^3 K_i^4 I_i^4$$
$$+ 0.065\,98 s_i^4 K_i^5 I_i^5 - b_i K_i \qquad (11.38)$$

where $K_i = 1/k_i$ and we have replaced s_E by s_i for element i.

One thing worth noting about this polynomial is that other than the background term which is negative, all the other terms are positive, which means it is a very well-behaved polynomial. Neither self-absorption nor the polynomial will go off and do funny wobbles or big side trips the way free polynomials can. However, it does not matter how many orders the polynomial is expanded into, the higher order terms always remain positive. When self-absorption is small it is possible to start with a first or second order and then include higher orders as s_i or $c_i q_M$ increases.

The most noticeable difference between the two models, *i.e.* between the simple, single emitting/absorbing region (exponential) model and the more complex, double region model, occurs at severe self-absorption. Since self-reversal cannot occur in a homogeneous (single region) plasma, the exponential model tends to underestimate severe self-absorption, see Figures 11.19 and 11.20.

For analysis, however, it is better to avoid situations of severe self-absorption because the measured signal becomes less and less sensitive to increasing $c_i q_M$ values. The solution is either to choose another spectral line of the element less subject to self-absorption or to reduce the sputtering rate by choosing softer plasma conditions, such as lower power. For less severe self-absorption, Figure 11.19 suggests that the exponential model should be adequate for most GDOES analysis.

Background Signal

The background is a difficult term to deal with in a general way, and, possibly because of this, it is the term least dealt with in the literature.[58,59] Background signals and inter-element effects (which are one particularly bothersome part of the background) are a complex source of problems for GDOES, as indeed they are for all emission spectroscopies. Background signals are especially annoying in GDOES depth profiles.

When using GDOES for bulk analysis, background signals are first noticed as the intercepts on the calibration curves. These intercepts, through their noise level, largely determine the detection limits for each element. Inter-element effects (*i.e.* interferences from neighbouring emission lines from argon or from the other elements in the sample) also act to increase the scatter on the calibration curves. In depth profiling, background and inter-element effects are even more serious: the background signal limits the dynamic range which can be measured and therefore displayed for

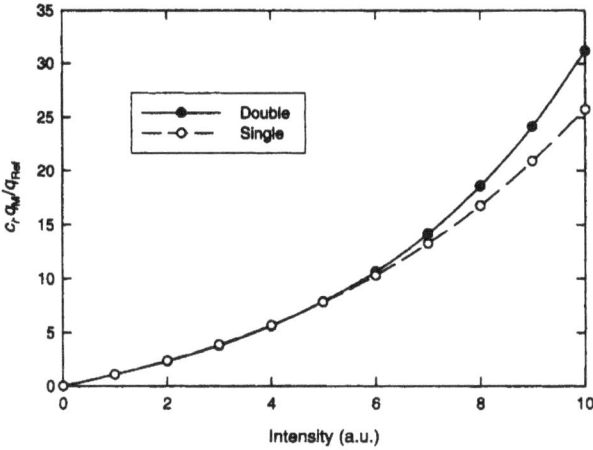

Figure 11.19   *Illustrating difference between the two self-absorption models*

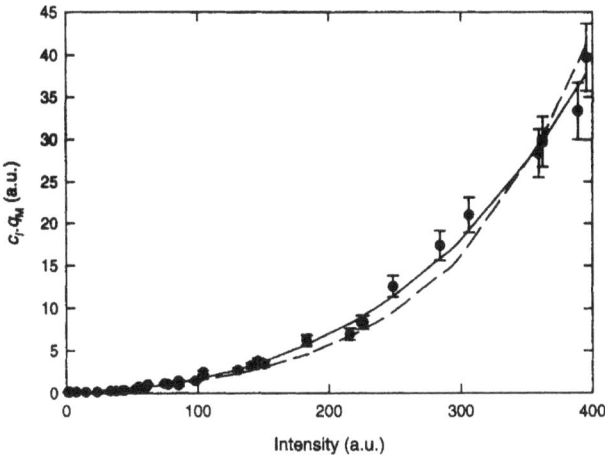

Figure 11.20   *Self-absorption for Zn 213 nm, data from Zn in a wide range of Al–Zn alloys.*[56]
*The solid line is obtained from Equation 11.38 and the dotted line is obtained
from the exponential model*

each element in the profile, and changes in the background signal in going from
one matrix to another during the depth profile can be confused with real changes in
composition.

The background term contains at least five separate components:[59] a constant
component from photomultiplier dark current and other instrumental noise sources;
continuum and line components from the argon plasma and continuum and line components from the sputtered matrix. The argon components might be expected to vary
with the argon pressure and with the power applied to the plasma, while the matrix
components might be expected to vary with the elemental sputtering rate and emission
yield.

Table 11.3 *Measured and calculated densities for a range of metals*

Alloy type	Content	Measured density[61–63] (kg m^{-3})	Calculated density (kg m^{-3})	Error (%)
Brass	Cu 70 Zn 30	8550	8323	−2.7
Cast iron	Fe 92 C 4 Mn 2	7697	7298	−5.2
Cast iron alloy	Fe 88 Ni 3 C 3 Cu 2	7743	7467	−3.6
Cast iron alloy	Fe 83 Si 4 Ni 3 C 2	7555	6979	−7.6
Copper–nickel	Cu 67 Ni 30 Fe 3	8900	8910	0.1
Copper–nickel	Cu 62 Ni 25 Zn 13	8820	8658	−1.8
Nickel high alloy	Ni 61 Cr 23 Fe 15	8110	8016	−1.2
Nickel high alloy	Ni 40 Fe 40 Cr 18	7910	7712	−2.5
Titanium alloy	Ti 83 Sn 11 Mo 4	4860	4725	−2.8
Steel high alloy	Fe 75 Cr 12 Ni 12	8010	7864	−1.8
Steel high alloy	Fe 69 Cr 30	7900	7596	−3.8
Steel high alloy	Cr 40 Fe 36 Ni 20	8020	7631	−4.9
Zinc–aluminium	Zn 90 Al 10	6116	6132	0.3
Zinc–aluminium	Zn 50 Al 50	3902	3918	0.4
Zinc–aluminium	Zn 10 Al 90	2878	2879	0.0

Note: Cast iron densities are measured by Archimedes' method.

Density

To convert the calculated sputtered mass into depth during compositional depth profiling, it is necessary to estimate the density of the sample or coating from the chemical composition at each measurement point. An effective way to do this for metals is to calculate the unknown density ρ from the densities of pure materials, assuming constant atomic volume, *i.e.*[60]:

$$\frac{1}{\rho} = \sum_i \frac{c_i}{\rho_i} \tag{11.39}$$

where c_i is the mass % of element i and ρ_i is the density of pure element i. Generally, this approach works well for metal alloys, typically giving results within a few per cent of measured values (see Table 11.3). It seems to work extremely well for zinc–aluminium alloys and less well for high-alloy steels and cast iron. The average error for the samples shown in the table is 2.6%.

Non-metals, in particular oxides, do not fair so well, the error being up to 40%. The assumption of constant volumes clearly does not work for these materials. The main reason lies in the nature of the chemical bonds: the stronger the bond the closer the atoms are drawn together and the smaller the atomic volumes; and some of the strongest bonds are in oxides. A secondary reason is the arrangement of atoms, *i.e.* the crystal structure, since a different structure may have a different number of atoms per unit volume.

Correction for Oxides and Nitrides

The constant atomic volume equation is not accurate for compounds with strong covalent bonds, such as oxides and nitrides, which reduce the apparent size of the atoms and hence increase the density.

Of course, we know from the analysis whether O and N or other covalent bond-forming elements are present. So, we can assume that if they are present, then they will form such bonds, and they will form them preferentially with those elements that give the lowest energy. We can then simply look up a table of densities for known oxides and nitrides and other compounds. To determine the density we then sum the specific densities (inverse densities) of the metals, oxides, nitrides, *etc.* into the density equation:

$$\frac{1}{\rho} = \left\{ \sum_i \frac{c_i'}{\rho_i'} \right\}_{\text{metals}} + \left\{ \sum_i \frac{c_i''}{\rho_i''} \right\}_{\text{oxides}} + \left\{ \sum_i \frac{c_i'''}{\rho_i'''} \right\}_{\text{nitrides}} + \cdots \quad (11.40)$$

where c_i' is the mass % of element i present as a metal (or isolated atom), c_i'' is the mass % of element i present as an oxide, c_i''' is the mass % of element i present as a nitride, and ρ_i' is the density of pure element i, ρ_i'' is the density of the preferred oxide of element i and ρ_i''' is the density of the preferred nitride of element.

To illustrate how the density calculation works for a pure Al_2O_3 sample, we already know that the composition is 53% Al and 47% O (this information is obtained from the y-axis of the compositional depth profile). Without an oxide correction, *i.e.* using Equation 11.40 with the densities for pure Al of 2700 kg m^{-3} and pure O of 3000 kg m^{-3} (a value chosen to give sensible results for a range of materials), we would calculate a sample density of 2830 kg m^{-3}, much lower than the correct density of 3980 kg m^{-3}. This would then lead to an unacceptable increase in the calculated depth. But with the oxide correction, we know that there is 47% O (from the y-axis) and that the lowest energy element present to form an oxide is Al (in this simple example, it is the only other element present), we identify from a table that the available quantity of O requires 53% Al to form Al_2O_3; look up the specific density for Al_2O_3, find there is no more Al unaccounted for and no other elements, and calculate a correct sample density of 3980 kg m^{-3}. This density is 40% greater than that calculated with the constant atomic volume equation and means that the new calculated depth will be about 30% less than the previous calculation.

References

1. M. Bouchacourt and F. Schwoehrer, in *Glow Discharge Optical Emission Spectrometry*, D.G. Jones and A. Bengtson (eds), John Wiley & Sons, Chichester, 1997, 51–3.
2. G.F. Weston, *Cold Cathode Glow Discharge Tubes*, Iliffe Books, London, 1968, 69, 98.
3. B. Chapman, *Glow Discharge Processes*, John Wiley & Sons, New York, 1981, 78, 122.
4. A. Bogaerts and R. Gijbels, *Spectrochim. Acta B*, 1998, **53**, 1.
5. K. Suzuki, *CAMP-ISIJ*, 1988, 1, 1619.
6. N.P. Ferreira and H.G.C. Human, *Spectrochim. Acta*, 1981, **36B**, 215.
7. A. Fiala, L.C. Pitchford and J.P. Bœuf, *Phys. Rev. E*, 1994, **49**, 5607.
8. A. Bogaerts, R. Gijbels and W.J. Goedheer, *Anal. Chem.*, 1996, **355**, 853.
9. P. Belenguer and L.C. Pitchford, *J. Anal. Atom. Spectrom.*, 2001, **16**, 1.
10. N. Matsunami, Y. Yamamura, Y. Itikawa, N. Itoh, Y. Kazumata, S. Miyagawa, K. Morita, R. Shimizu and H. Tawara, *Atom. Data Nucl. Data Tables*, 1984, **31**, 1.
11. A. Bogaerts and R. Gijbels, *J. Anal. Atom. Spectrom.*, 1998, **13**, 721.
12. A. Bogaerts and R. Gijbels, *Spectrochim. Acta B*, 1997, **52**, 553.

13. A. Bogaerts, L. Wilken, V. Hoffmann, R. Gijbels and K. Wetzig, *Spectrochim. Acta B*, 2001, **56**, 551.
14. A. Bogaerts, R. Gijbels and J. Vlcek, *Spectrochim. Acta B*, 1998, **53**, 1517.
15. A. Bogaerts and R. Gijbels, *Spectrochim. Acta B*, 1998, **53**, 437.
16. A. Bogaerts and R. Gijbels, *Spectrochim. Acta B*, 2000, **55**, 279.
17. A. Bogaerts, R. Gijbels and R.J. Carman, *Spectrochim. Acta B*, 1998, **53**, 1679.
18. R. Payling, P. Chapon, O. Bonnot, P. Belenguer, P. Guillot, L.C. Pitchford, L. Therese, J. Michler, and M. Aeberhard, *ISIJ Int.*, 2002, **42**, S101.
19. Ph. Belenguer, Ph. Guillot and L. Therese, CPAT-UPS, France, private communication, 2002.
20. L. Therese, P. Guillot, P. Belenguer, L.C. Pitchford, R. Payling, O.Bonnot and P. Chapon, *Plasma Winter Conf.*, 2002.
21. A. Bogaerts, R. Gijbels and W. Goedheer, *J. Anal. Atom. Spectrom.*, 2001, **16**, 750.
22. A. Bogaerts, L. Wilken, V. Hoffmann, R. Gijbels and K. Wetzig, *Spectrochim. Acta B*, 2002, **57**, 109.
23. A. Bogaerts and R. Gijbels, *J. Anal. Atom. Spectrom.*, 2000, 15, 1191.
24. J.P. Boeuf and P. Belenguer, in *Nonequilibrium Processes in Partially Ionized Gases*, M. Capitelli and J.N. Bardsley (eds), Plenum Press, New York, 1990, 220, 155.
25. A. Bogaerts, R. Gijbels and W. Goedheer, *Spectrochim. Acta B*, 1999, **54**, 1335.
26. A. Bengtson and S. Hänström, in *Proc. 5th Int. Conf. on Progress in Analytical Chemistry in the Steel and Metals Industries*, R. Tomellini (ed), European Communities, Luxembourg, 1999, 47–54.
27. V.-D. Hodoroaba, V. Hoffmann, E.B.M. Steers and K. Wetzig, *J. Anal. Atom. Spectrom.*, 2000, **15**, 1075.
28. R. Payling, M. Aeberhard and D. Delfosse, *J. Anal. Atom. Spectrom.*, 2001, **16**, 50.
29. A. Bogaerts, *J. Anal. Atom. Spectrom.*, 2002, **17**, 768.
30. R.A. Baragiola, E.V. Alonso, J. Ferron and A. Oliva Florio, *Surf. Sci.*, 1979, **90**, 915.
31. L.M. Kishinevsky, *Radiat. Eff.*, 1973, **19**, 23.
32. L. Ohannessian, PhD Thesis, Universite Claude Bernard, Lyon, France, 1986.
33. H. Hocquaux, in *Glow Discharge Spectroscopies*, R.K. Marcus (ed), Plenum Press, New York, 1993, 351.
34. R.A. Baragiola, E.V. Alonso, J. Ferron and A. Oliva Florio, *Surf. Sci.*, 1979, **90**, 915.
35. L.M. Kishinevsky, *Radiat. Eff.*, 1973, **19**, 23.
36. R.A. Baragiola, private communication, 2000.
37. R. Payling, 2000, www.glow-discharge.com/secondary_electron_yield.htm
38. L. Wilkin, V. Hoffmann and K. Wetzig, *Proc. Final General Meeting, EC Thematic Network on GDS for Spectrochem. Anal.*, Wiener Neustadt, Austria, 2002.
39. J. Stark, *Z. Elektrochem.*, 1909, **15**, 509.
40. M. Kaminsky, *Atomic and Ionic Impact Phenomena on Metal Surfaces*, Springer-Verlag, Berlin, 1965.
41. R. Payling and D.G. Jones, *Surf. Interface Anal.*, 1993, **20**, 787.
42. B.V. King, in *Glow Discharge Optical Emission Spectrometry*, R. Payling, D.G. Jones and A. Bengtson (eds), John Wiley & Sons, Chichester, 1997, 245–53.
43. A. Benninghoven, F.G. Rüdenauer and H.W. Werner, *Secondary Ion Mass Spectrometry*, John Wiley & Sons, New York, 1987.
44. R. Payling, *Surf. Interface Anal.*, 1994, **21**, 791.
45. R. Payling, in *Glow Discharge Optical Emission Spectrometry*, R. Payling, D.G. Jones and A. Bengtson (eds), John Wiley & Sons, Chichester, 1997, 260, 267.
46. R. Payling, M. Aeberhard, J. Michler, C. Authier, P. Chapon, T. Nelis and L. Pitchford, *Surf. Interface Anal.*, 2003, **35**, 334.

47. R. Payling, D.G. Jones and S.A. Gower, *Surf. Interface Anal.*, 1993, **20**, 959.
48. R. Payling, D.G. Jones and S.A Gower, *Surf. Interface Anal.*, 1995, **23**, 1.
49. Y. Ishibashi and Y. Yoshioka, *Trans. ISIJ*, 1988, **28**, 773.
50. M.K. Levy, D. Serxner, A.D. Angstadt, R.L. Smith and K.R. Hess, *Spectrochim. Acta*, 1991, **46B**, 253.
51. R. Payling, *Surf. Interface Anal.*, 1995, **23**, 12.
52. A. Bengtson and M. Lundholm, *J. Anal. Atom. Spectrom.*, 1988, **3**, 879.
53. K. Wagatsuma, in *Glow Discharge Optical Emission Spectrometry*, R. Payling, D.G. Jones and A. Bengtson (eds), John Wiley & Sons, Chichester, 1997, 342.
54. R. Payling and P.L. Larkins, *Optical Emission Lines of the Elements*, John Wiley & Sons, Chichester, 2000.
55. M.R. Winchester, private communication, 2002.
56. R. Payling, M.S. Marychurch and A. Dixon, in *Glow Discharge Optical Emission Spectrometry*, R. Payling, D.G. Jones and A. Bengtson (eds.), John Wiley, Chichester, 1997, 376–91.
57. R. Payling, *Spectroscopy*, 1998, **13**, 36.
58. A. Bengtson and A. Eklund, *Comm. Eur. Communities Rept.*, **EUR 14113**, *Prog. Anal. Chem. Iron Steel Ind.*, 1992, 43–8.
59. R. Payling, N.V. Brown and S.A. Gower, *J. Anal. Atom. Spectrom.*, 1994, **9**, 363.
60. R. Payling, in *Glow Discharge Optical Emission Spectrometry*, R. Payling, D.G. Jones and A. Bengtson (eds), John Wiley & Sons, Chichester, 1997, 287–91.
61. M. Schierling and R. Harris, *Can. Metall. Q.*, 1992, **31**, 241.
62. E.A. Brandes and G.B. Brook (eds), *Smithells Metals Reference Book*, Butterworth, London, 1992, 3-1–3-2, 14-1–14-2, 18-4.
63. J.A. Dean (ed), *Lange's Handbook of Chemistry*, McGraw-Hill, New York, 1985, 4-14–4-133, 7-82–7-279.

Line Selection

The selection of which lines to use is a key part of GDOES, both in the design of instruments and in analysis. If an instrument has a polychromator with fixed exit slits, the choice has largely been made by the instrument manufacturer during the construction of the instrument. Such polychromators have only one or two lines per element; though, as the needs of the laboratory change, PMTs may be shifted to other wavelengths or new PMTs may be added to provide additional lines or elements. If the instrument has a scanning monochromator or simultaneous solid-state spectrometer, then there is an enormous choice of lines, either as additional lines to complement the polychromator (if both are present) or in stand-alone analysis. The use of solid-state detectors in GD instruments will increase in the future, as it did in ICP and Spark OES. The use of several spectral lines for each element will probably also gain in interest. Some knowledge of line selection is therefore essential.

1 Wavelength Convention

Radiation is emitted from atoms in discrete quanta, called photons, whose energy ΔE is equal to the difference in energy between the initial i and final f stationary states of the atom, and is proportional to frequency, ν:

$$\Delta E = E_i - E_f = h\nu \tag{12.1}$$

In optical spectroscopy we normally use wavelength rather than frequency; hence the observed wavelength in vacuum, λ_{vac}, is:

$$\lambda_{vac} = \frac{c}{\nu} = \frac{hc}{\Delta E} \tag{12.2}$$

where c is the speed of light in vacuum.

The frequency of the EM field associated with the photon does not change when it travels from one medium to another but both speed and wavelength do change. The refractive index of a medium is defined as the ratio of the speed of light in vacuum to the speed of light in the medium. The difference in wavelength in air and vacuum is therefore given by:

$$\lambda_{vac} - \lambda_{air} = (n_{air} - 1)\lambda_{air} \tag{12.3}$$

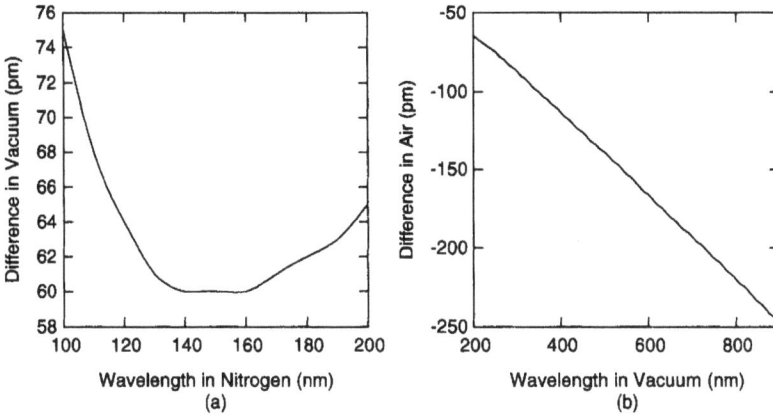

Figure 12.1 *Conversions between wavelengths in air/nitrogen and vacuum, add difference: (a) below 200 nm to convert from nitrogen to vacuum, (b) above 200 nm to convert from vacuum to air*

where n_{air} is the refractive index of air. Since n_{air} is slightly greater than 1, λ_{air} is slighter less than λ_{vac}. The refractive index of air varies with the composition of the air and its density, *i.e.* it varies with the purity of the air and its pressure and temperature. The refractive index of nitrogen is very close to that of air, except that high-purity nitrogen transmits wavelengths below 180 nm that air will not.

By convention, the emission lines above 200 nm are labelled by their values in air. This is done even when the wavelengths are measured in vacuum and then converted to air values. Below 200 nm, emission lines are labelled by their values in vacuum, even if measured in nitrogen and converted to vacuum. Hence Al 396.152 is an air wavelength, while C 156.140 is a vacuum wavelength. All the wavelengths shown in this book are in nanometres. For nitrogen-filled spectrometers, measured values below 200 nm are converted from nitrogen to vacuum, *i.e.* by adding the difference shown in Figure 12.1a. For vacuum spectrometers, measured values above 200 nm are converted from vacuum to air, *i.e.* by adding the difference shown in Figure 12.1b.[1]

2   GDOES Spectra

Theoretical spectra are available for atoms and their first ionic states for virtually all elements in the periodic table.[1] Such spectra assume that the emitting atoms are in local thermodynamic equilibrium (LTE). This assumption is necessary to estimate the populations of atoms and ions in their various excited states, according to Boltzmann statistics. The wavelength of the emission line depends on the difference in energy between the initial (excited) and final (de-excited) states of the atom and in analytical plasmas is largely unaffected by the plasma. The intensity of an emission line (neglecting self-absorption), on the other hand, is proportional to the number of atoms in the excited state. The GD plasma is not in LTE. This means that while the wavelengths of the emission lines will be nearly the same as those from other plasmas (neglecting small pressure effects), the intensities, and especially the ratio of neutral to ionic lines,

Table 12.1 *Most intense lines for Al*

Wavelength (nm)	Lower energy (eV)	Intensity (calculated, a.u.)	Intensity (measured, a.u.)
396.152	0.014	870 000	870 000
394.401	0.000	435 000	750 000
309.271	0.014	261 000	335 000
308.215	0.000	148 000	245 000
309.284	0.014	28 300	240 000
256.798	0.014	19 000	33 500
266.039	0.000	56 000	33 500

may be quite different. Hence in GDOES the best emission lines for analysis may be different from other optical emission techniques, such as Spark source OES and ICP-OES.

For instruments with monochromators or simultaneous solid-state spectrometers, it is relatively easy to record a large section of the optical spectrum. For pure, or nearly pure, single-element materials it is therefore relatively straightforward to determine the strongest lines of that element, by comparing observed lines with tables of known lines, *e.g.* Harrison,[2] Striganov and Sventitskii,[3] and Payling and Larkins.[1] An example for Al is shown in Table 12.1. Here, the most intense lines in the GD spectrum were selected and then identified by comparison with the calculated wavelengths and intensities using the program supplied on the CD-ROM by Payling and Larkins.[1] The calculated intensities were normalised to match the measured intensity for Al 396.152.

The main complication is the presence of Ar lines. A rapid way to identify Ar lines, suggested by Bonnot,[4] is to record spectra from two different pure materials (notably copper and silicon, see Figure 12.2), remove a constant background and then multiply the two spectra. Since Ar lines will be in both spectra the multiplied spectra will highlight the argon lines. The resulting Ar spectrum, being the square root of the multiplied spectrum, is shown in Figure 12.3 and some of the strongest lines are listed in Table 12.2. Many of the lines in the Si spectrum are then seen to be Ar lines.

It is much more difficult to determine the most intense lines for elements that are not available as suitable single-element materials. One way is to take two or three samples, preferably in a similar matrix, with quite different compositions of the element of interest and overlay their spectra. The lines of interest should show intensities nearly proportional to their $c_i q_M / q_{Ref}$ products, *i.e.* to the product of the element's composition and the relative sputtering rate of the matrix. If the samples have the same matrix, then the intensities should be nearly proportional to the compositions.

3 Best Analytical Lines

The best emission lines for analytical use should be judged on the following criteria:[5,6]

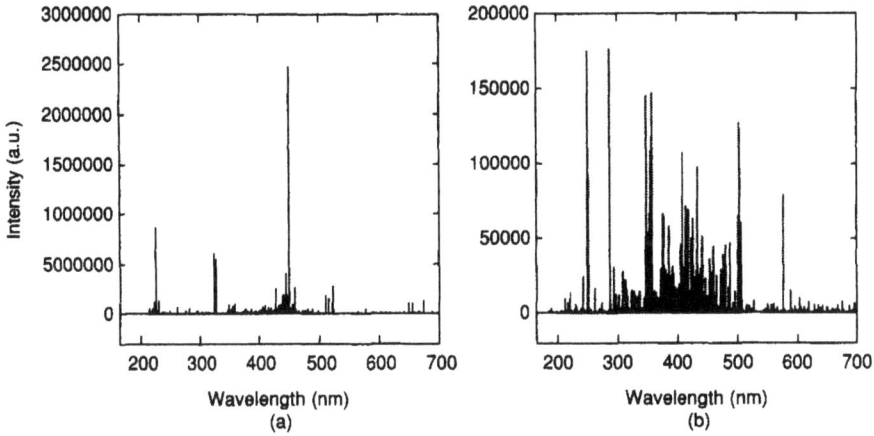

Figure 12.2 *Spectrum from: (a) nearly pure copper, (b) high-purity silicon*

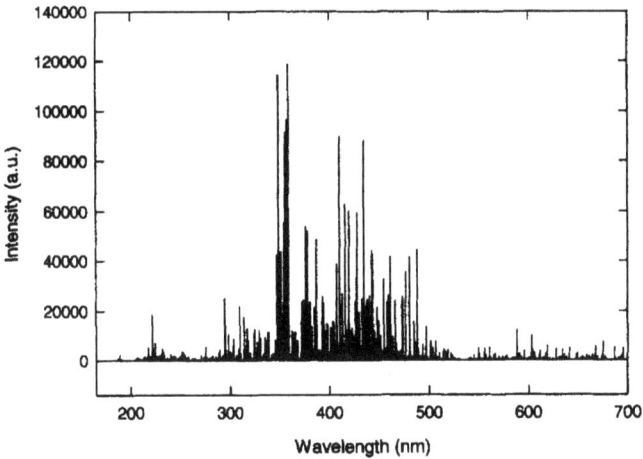

Figure 12.3 *Multiplied spectrum of Ar*

- high sensitivity (improves detection limit)
- high signal-to-background ratio (reduces the background equivalent composition)
- high signal-to-noise ratio (improves detection limit)
- low spectral interferences (reduces scatter on calibration curves)
- low self-absorption (provides linear calibration curves)
- good reproducibility for the desired composition range
- small effect of trace gases on the emission yield

Such characteristics would lead to linear calibration curves with little scatter and low detection limits. For some elements, *e.g.* B, it is relatively easy to fulfil all or many of these criteria, but for other elements, *e.g.* Al, there are few suitable lines to choose

Table 12.2   *Most intense lines*
observed for Ar

Wavelength (nm)	Intensity (a.u.)
358.844	20 090
349.154	19 120
357.662	15 710
355.951	15 520
410.391	15 190
434.806	14 160
358.241	12 040
415.859	10 400
420.068	10 150
427.752	9 960
354.584	9 950
376.527	9 270
378.084	8 650

from and most of the intense lines show self-absorption. In some cases, more than one line can be chosen, to cover different composition ranges (*e.g.* because of limited sensitivity or self-absorption) or for use in different matrices (*e.g.* because of spectral interferences).

The key parameters (refer to Figure 12.4) for assessing the suitability of an emission line are the following:

- intensity, S
- background, B
- signal-to-background ratio, SBR = S/B
- standard deviation of the signal, SDS, normally measured by repeating ten times on the peak position
- relative standard deviation of the signal, RSDS = SDS/S, usually expressed as %
- standard deviation of the background, SDB, normally measured from a scan (as in Figure 12.4) or by repeating ten times away from the peak
- relative standard deviation of the background, RSDB = SDB/B, usually expressed as %
- signal-to-noise ratio, SNR = S/SDB
- detection limit, based on signal (SBR-RSDB approach), DL = $0.03 \times$ RSDB $\times c_o$/SBR,[7] for three standard deviations, where c_o is the content of the element producing the signal

It is even sensible, though not yet a common practice, to use several spectral lines simultaneously, reducing the sometimes unexpected effects of interference and matrix dependence of response. Different lines could be ratioed to different internal standards, to reduce the effect of non-linear reference lines. The use of several lines

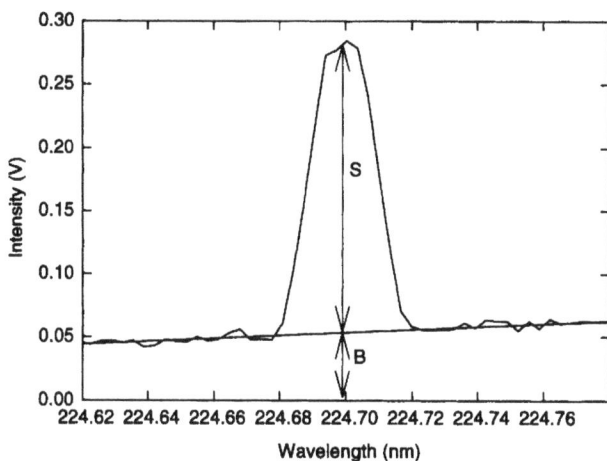

Figure 12.4 *Illustration of signal S and background B for 13 ppm Cu 224.700 in steel NBS 1765A*

also improves statistics, as more information is drawn from the spectrum emitted by the plasma.

4 Interferences

Spectral interferences are not usually a major problem in GDOES, but they do occur, in particular when spectrometers with low resolving power are used. The proper identification of spectral interferences and correction in calibration can add significantly to the accuracy of analysis when they are present. On the other hand, wrongly correcting for a spectral interference simply because it makes the calibration curve look better when no interference can lead to highly inaccurate analysis. The lesson (worth repeating here) is: only include corrections for a spectral interference when one is sure it is present and has verified that the correction is working correctly on validation samples.

Determination of the presence of a spectral interference in GDOES can be a difficult task. Unlike ICP-OES, where it is possible to prepare comparable samples with and without a suspected interferant, in GDOES the analyst is normally compelled to use the solid samples at hand. It is usually difficult to find two solid samples of the same material type, one with and one without the suspect element.

Spectral interference can arise from overlapping lines from other elements (see Figure 12.5) or from changes to the background signal due to scattered light inside the spectrometer. Generally, the seriousness of the interference increases with increasing composition of the interfering element. Major elements are therefore the elements of most concern; it is extremely unlikely that a trace element would cause a significant interference. If the interfering line does not completely overlap the analytical line, it may be possible to see the interfering line directly by scanning around the analytical line. In a polychromator this can be done by scanning the primary slit. But often the

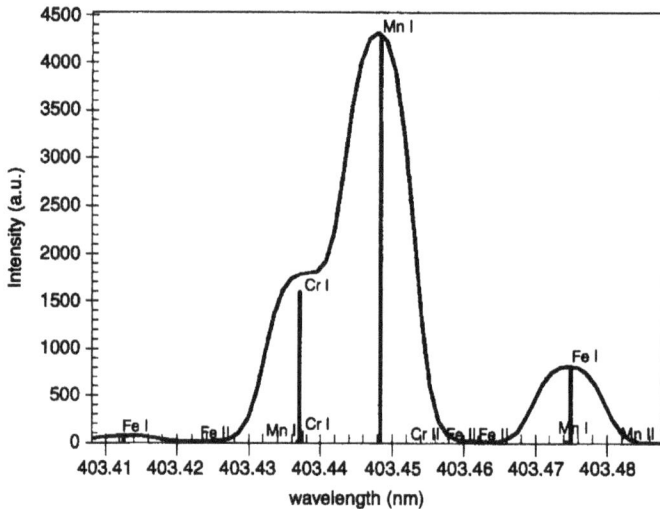

Figure 12.5 *Calculated spectrum, with a spectral resolution of 9 pm, showing Cr 403.437 interfering with Mn 403.448 and with Fe 403.475 nearby*

interfering line is too close to resolve. Then the main evidence for its presence is scatter in the calibration curve. But to be sure that it is really present and not just due to noise, it is best to check suspected interferences with tables of known interferences or with tables of emission lines. A list of known and possible interferences for the lines commonly used in GDOES is provided in Appendix B.

In calibration, the operator must identify a suspected interfering element (it is dangerous to do this automatically), and the magnitude of the interference is then determined by regression, assuming that the interference is linear with the intensity of a known line of the suspected interferant. The success of this process will depend on having a sufficient number of calibration samples with varying contents of the interfering element.

5 Linearity, Self-absorption

The main cause of non-linearity in calibration curves is self-absorption. Other possibilities are systematic changes in sputtering rates or emission yields. Only resonance lines or near-resonance lines are known to show self-absorption in GDOES. These lines can be identified from the energy of the final state, being less than 0.25 eV (2000 wavenumbers). All the Al lines listed in Table 12.1, therefore, are resonance or near-resonance lines and subject to self-absorption.

Spectral lines that are used as internal standards in relative methods must have a linear response; otherwise, the approach leads to inaccurate results. Similarly, most software requires that lines used for interference corrections be linear.

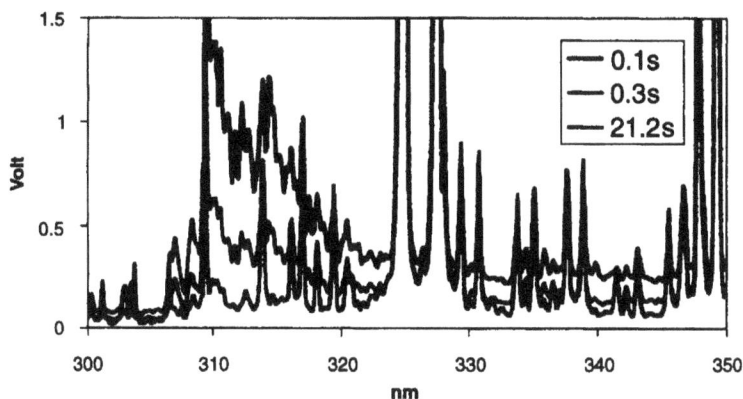

Figure 12.6 *Time-resolved spectra from a Cu surface slightly contaminated by a finger-print, showing lines for Cu, V, Zn and Bi lus molecular bands, probably including OH*
(Reproduced with permission from A. Bengtson and S. Hänström, *Proc. 2002 Winter Conf. Plasma Spectrochemistry*, Scottsdale, USA, 6–12 Jan., 50–51, 2002)

6 Molecular Bands

There is renewed interest in molecular bands, especially at the extreme surface, where OH bands have been detected.[8] Such molecular bands are interesting in themselves, if they can be linked to sample chemistry, see, for example, Figure 12.6. They are also interesting if they cause spectral interferences with line spectra used to monitor elements. Such interferences may explain the unexpected signals often seen for some elements at extreme surfaces.

Little information is currently available on the presence of molecular bands in GDOES spectra. Four molecular bands were reported for N_2 in the region 330.3–338.6 nm, using a Fourier transform spectrometer (FTS), when nitrogen gas was deliberately added to the argon gas.[9] The increased use of solid-state spectrometers in the future should facilitate the study of such bands.

An important source of information on molecular emission spectra in the range from UV to near-IR of interest in GDOES is the work of Pearse and Gaydon.[10] Some of their data for OH and N_2 are shown in Table 12.3. OH is present on the surface of most materials examined by GDOES. N_2 may be present if air leaks past the O-ring seal into the plasma. Other binary molecules (including ionised species) of potential interest are C_2, CH, CN, CO, CS, NH, NO, O_2, SO and metal hydrides, which result from various possible chemical reactions in the plasma involving hydrocarbons, water or air. The O_2 molecule is not a strong emitter in low-pressure discharges.[10]

Molecular spectra are characterised by peaks in intensity, called heads, which decrease (degrade) in intensity due to either shorter or longer wavelengths depending on the nature of vibrations in the particular molecule. The recommended procedure for identifying a molecular spectrum is to find a strong (persistent) head, verify that it degrades in the right direction and then look for several other heads. For example,

Table 12.3 *Persistent band heads of OH and N_2 between 200 nm and 500 nm[10]*

Wavelength (nm)	Description	Molecule
281.13	R, DCD, wv	OH
297.68	V, CT	N_2
306.36	R, DCD, wv	OH
315.93	V, CT	N_2
337.13	V, CT	N_2
357.69	V, CT	N_2
358.21	V	N_2^+
380.49	V, CT	N_2
391.44	V	N_2^+
405.94	V, CT	N_2
427.81	V	N_2^+
434.36	V, CT	N_2

V means degraded (decreasing) to shorter wavelengths (violet); R means degraded to longer wavelengths (red); DCD means double head and each component is a close double head; T is triple head; CT is close triple head; and wv means with an additional weaker head to shorter wavelengths.

if we examine the spectrum in Figure 12.6 we find a head at about 308 nm. This is close to a persistent head for OH (see Table 12.3). As expected for OH, it degrades to the red. The OH head is labelled DCD, which means that it is in fact four main heads, lying between 306 and 309 nm. There are several weaker heads near 302, 312, 319 and 343 nm. To verify that it really is OH, it would be worthwhile to also look for the other persistent head at 281 nm.

References

1. R. Payling and P.L. Larkins, *Optical Emission Lines of the Elements*, John Wiley & Sons, Chichester, 2000.
2. G.R. Harrison, *Wavelength Tables with Intensities in Arc, Spark, or Discharge Tube of more than 100,000 Spectrum Lines*, The M.I.T. Press, Cambridge, MA, 1969.
3. A.R. Striganov and N.S. Sventitskii, *Tables of Spectral Lines of Neutral and Ionized Atoms*, IFI/Plenum Press, New York, 1968.
4. O. Bonnot, private communication, 1999.
5. T.R. Harville and R.K. Marcus, *Anal. Chem.*, 1993, **65**, 3636.
6. T.R. Harville, in *Glow Discharge Optical Emission Spectrometry*, R. Payling, D.G. Jones and A. Bengtson (eds), John Wiley & Sons, Chichester, 1997, 360.
7. P.W.J.M. Boumans, in *Glow Discharge Optical Emission Spectrometry*, R. Payling, D.G. Jones and A. Bengtson (eds), John Wiley & Sons, Chichester, 1997, 440–51.
8. A. Bengtson and S. Hänström, *Proc. 2002 Winter Conf. Plasma Spectrochemistry*, Scottsdale, USA, 6–12 Jan., 50–1, 2002.
9. P. Smid, E.B.M. Steers and Z. Weiss, *Proc. EC Thematic Network on GDS General Meeting*, Wiener Neustadt, 3–6 Mar., 2002.
10. R.W.B. Pearse and A.G. Gaydon, *The Identification of Molecular Spectra* 4th edn, Chapman and Hall, New York, 1976.

CHAPTER 13

Troubleshooting

1 Diagnosis

Much of the detailed maintenance and repair of GDOES instruments, especially inside optical spectrometers, is best left to service engineers. But there are some common problems that the operator can deal with, and diagnosing the problem before the service engineer arrives can greatly simplify the task for the engineer and speed up the repair work. It may mean, for example, that the engineer arrives with the right spare parts for the job.

The first task in diagnosing a problem is to identify which part of the instrument needs attention. The instrument comprises the following features.

Sample

- porosity
- surface finish
- temperature stability

The sample forms the cathode of the GD source; it is therefore an important part of the instrument and can affect the performance of the instrument. Porous samples will allow air to enter the source. If the air ingress is not too high, it may still be possible to ignite the plasma but the resulting plasma will be quite different from the typical argon plasma and it will be difficult to compare results with other samples. If one suspects that the sample may be porous, then O and N (if available) should be included in the analysis; there will be unusually high O and N signals. Manufacturers can provide special sample holders for porous samples. Another option is to coat the outside of the sample with a non-porous coating, e.g. gold or nail polish.

Air ingress can also come from very rough sample surfaces that do not seal properly on the O-ring. Scratches on the sample surface can also allow air past the O-ring.

Low-temperature metals may evaporate rather than sputter if the power is too high. This can give rise to unstable depth profiles. Likewise, some glasses may shatter from thermal shock; normally this will allow so much air into the source that no plasma is possible.

GD Source

- anode
- window
- O-rings
- sample cooling circuit

The anode needs constant care and attention by the operator, mainly to ensure it is kept clean. Normally, this is done routinely between analyses by automatic or manual reaming. Occasionally, for copper anodes, when they become black from excess use, it is advisable to remove the anode and immerse it for a few minutes in dilute hydrochloric acid, then rinse and dry before replacing it, and finally run a sample for a few minutes to give the anode enough time to settle in before continuing analysis.

It is also advisable to check the anode-to-sample gap occasionally to see if it is still within specifications (typically between 0.1 mm and 0.25 mm). Some manufacturers suggest that the anode be replaced at specified periods. Others suggest that it be replaced only when it is damaged or the anode-to-sample gap is too large.

The window needs cleaning occasionally, see discussion in Chapter 1. Likewise the cathode block, *i.e.* the part of the source containing the O-ring seal, needs occasional cleaning.

The O-ring sealing the sample to the source is a critical component. It should be kept clean and lint-free. Occasionally, O-rings are damaged during removal, resulting in small cuts or cracks that are easily overlooked. O-rings can also be damaged by excessive sample heating, usually caused by air entering the plasma between the O-ring and sample surface. When overheated, O-rings may lose elasticity and shape. If the instrument is working but giving unusual results, then it may be worth changing the O-ring to see if it fixes the problem.

Occasionally, connections in the sample cooling system can leak. These should be fixed quickly because of the high voltages present in the source. The cooling fluid should be changed routinely, perhaps every 6 or 12 months. Contamination of the cooling fluid can lead to power losses with both RF and DC operations.

When replacing the cooling fluid, follow the manufacturer's instructions carefully. Highly deionised water and distilled water should be avoided. Some tap waters contain high levels of minerals that can cause severe power losses in RF and DC operations and should also be avoided. Normal laboratory-standard, deionised water is preferred. Corrosion inhibitors and biocides are sometimes added, though glycol is not recommended as it can cause power losses. Isopropyl alcohol or ethanol is sometimes used, up to 50% in deionised water. Before trying something new, it is worth checking first with the manufacturer and also checking to see if it causes power losses in the source. The easiest way to find this is to run a sample and see if the intensities change when the chemical is added. There should be no observable change.

Source Ancillary Components

- vacuum tubes, connectors, valves, gauge(s)
- pump(s)
- gas supplies

For the instrument to operate well, the GD source must be capable of reaching its normal (residual) vacuum and being supplied with high-purity argon at stable operating conditions. If you suspect some problem, usually because you cannot get the results you normally do, first check the residual vacuum. If the instrument allows it, place a smooth, flat sample on the source and note the vacuum the source reaches without argon. If this is high, it indicates either the pump is not working efficiently or there is a leak.

If the residual vacuum is normal, the problem may lie in the purity of the argon being supplied. First check whether the supply is normal. If this is true, then, if O and N are available on the instrument, run a sample low in these elements (*e.g.* pure copper) and check whether the O and N signals are both unusually high, this would indicate air is getting into the gas lines. If only O is high this indicates the argon is contaminated. If only N is high, and you have a nitrogen-purged spectrometer, then nitrogen may be getting past the window from the spectrometer into the source. Remove the window, check the window and the seal, and replace on the instrument.

Check the stability of the source by running a bulk sample (*e.g.* pure copper) and monitoring the source parameters, such as power, voltage, current and pressure. Unusual fluctuations in the electrical parameters should be noted to help the service engineer diagnose the cause of the problem. Unusual fluctuations in pressure may indicate a problem in gas flow, such as insufficient supply pressure or a fault in the flow controller.

Electronics

- system control
- communications

If part of the instrument does not function in response to software, then there may be a problem with the instrument's electronics. Usually, the site of the problem can be identified by built-in diagnostics or by speaking with a service engineer.

One should not trust the electronic values shown by an instrument blindly, especially as the instrument gets older. These values depend on components that may have changed, either during repairs or through ageing. An occasional check is recommended, for example, on the voltage, current, power and pressure values shown during DC or RF operation.

Optical Spectrometer(s)

- detectors
- vacuum or nitrogen-purge

Polychromators require periodic alignment of the primary slit. This is described in Chapter 1.

Normally, optical spectrometers should only be opened by trained engineers because of delicate components and the careful alignment of the spectrometer. But the operator should be aware of the spectrometer's performance, *e.g.* noting if a particular PMT no longer gives a signal or has an unusually high background or noisy signal.

Also, if the spectrometer is working at wavelengths below 180 nm, then it is either under vacuum or nitrogen-purged. Loss of signals below 180 nm indicates a problem with the cleanliness of the window or with the vacuum or nitrogen.

Computer

- hardware
- software
- cables
- printer

Computer hardware is generally very reliable. Most computer problems encountered in operating instruments, such as the software freezing, are therefore related to the computer's operating system. Usually rebooting will fix these, though it should not be necessary to do this often, for example, every day.

If it is necessary to reboot the computer often, then this usually indicates a more serious problem. Typically, this can happen if the hard disk is getting too full (less than 10% is free) or if the amount of RAM is insufficient. The obvious solutions to the hard disk problem are to clean out old files from the hard disk or to get a bigger disk. The RAM problem can arise from upgrades in software, either of the operating system or the instrument software, as these often require more RAM than older programs.

No software is perfect. If you find a 'bug', you should notify the manufacturer. Normally, manufacturers are only too happy to fix software problems. But you must provide sufficient information for their programmers to reproduce the problem. You should supply the version number of the software and the exact steps you followed leading up to the problem. If the problem is obvious on the screen, then you could do a screen dump and send this too.

2 Troubleshooting Guide

Symptom	Possible cause	Action
Source not starting	Short circuit: usually caused by dust on the anode	1. Clean anode and rerun sample. Can be a common problem, depending on the sample type. If it occurs often, wipe the anode face with a lint-free cloth before each analysis
Sample very hot	Air getting into source	1. Check sample for scratches or porosity 2. Check O-ring for cuts or loss of shape

Symptom	Possible cause	Action
Low sputtering rate	Air getting into source	1. Check sample for scratches or porosity 2. Check O-ring for cuts or loss of shape
High N and O signals	Air getting into source	1. Check sample for scratches or porosity 2. Check O-ring for cuts or loss of sample shape 3. Check argon supply
High O signal but low N signal	Contamination of argon	1. Check argon supply 2. If using bottles, try a different bottle
High N signal but low O signal	Nitrogen getting into source	1. If nitrogen-purged spectrometer, check window 2. Check purity of argon supply
Low signals below 180 nm	Dirty window	1. Clean window
	Poor vacuum in spectrometer (if vacuum spectrometer)	1. Check vacuum 2. Check vacuum pump 3. Check vacuum connections
	Insufficient nitrogen in spectrometer (if nitrogen-purged spectrometer)	1. Check nitrogen supply 2. Check gas connections
Intensities low or unstable	Spectrometer alignment	1. Run alignment procedure in software
High residual vacuum	Poor vacuum in source	1. Check vacuum pump(s) 2. Check window seal 3. Check O-rings 4. Check mounting of source by disassembling and reassembling 5. Check vacuum connections
Part of instrument does not respond	Electronics fault	1. Identify the location of the fault by discussing the symptoms with service engineer
Source unstable	Pressure fluctuation	1. Check argon supply pressure 2. Flow controller may need service

Symptom	Possible cause	Action
	Short circuit	1. Clean anode and rerun sample.
	Power cable not connected	1. Reconnect cable
Compositions change abruptly at interfaces in CDP	Calibration error	1. Check calibration curves proceed correctly near zero intensity
Compositions persist wrongly in CDP	Source parameters or calibration error	1. Check crater shape 2. Check calibration curves proceed correctly near zero intensity
Computer screen freezes often	Hard disk nearly full	1. Clean out unwanted files 2. Increase size of hard disk
	Insufficient RAM, after upgrade	1. Increase size of RAM
	Address conflict, after hardware change	1. Change settings in operating system, after consulting with service engineer

Further Reading

General Reading on Glow Discharges
B. Chapman, *Glow Discharge Processes*, John Wiley & Sons, New York, 1980.
G.F. Weston, *Cold Cathode Glow Discharge Tubes*, Iliffe Books, London, 1968.

Background Reading
R.K. Marcus and J.C. Broekaert (eds), *Glow Discharge Plasmas in Analytical Spectroscopy*, John Wiley & Sons, Chichester, 2003.
A. Bogaerts and R. Gijbels, Fundamental aspects and applications of glow discharge spectrometric techniques, *Spectrochim. Acta B*, 1998, **53**, 1–42.
R. Payling, D.G. Jones and A. Bengtson (eds), *Glow Discharge Optical Emission Spectrometry*, John Wiley & Sons, Chichester, 1997.
R.K. Marcus (ed), *Glow Discharge Spectroscopies*, Plenum Press, New York, 1993.

Theory
*A. Bogaerts and R. Gijbels, Comprehensive description of a Grimm-type glow discharge source used for optical emission spectrometry: a mathematical simulation, *Spectrochim. Acta B*, 1998, **53**, 437–62.
T.P. Softley, *Atomic Spectra*, Oxford Science Publications, Oxford, 1994.

Numerical Methods and Regression
W.H. Press, S.A. Teukolsky, W.T. Vetterling and B.P. Flannery, *Numerical Recipes in Fortran 77*, 2nd edn, Cambridge University Press, 1997; also in C^{++} and Pascal.
*J.O. Rawlings, S.G. Pantula and D.A. Dickey, *Applied Regression Analysis*, 2nd edn, Springer, Berlin, 1998.

Practice
EURACHEM/CITAC Guide, *Quantifying Uncertainty in Analytical Measurement*, 2nd edn, 2000, available for free download as a pdf file from the Internet.

Applications
R. Payling, Glow discharge optical emission spectrometry, *Spectroscopy*, 1998, **13**, 36–44.

Books and papers marked * are more advanced.

Appendix

Appendix A. Most Used Optical Emission Lines in GDOES

Element	State	Line (nm)	Upper energy (eV)	s_L $(\times 10^{-12}\ m^2)$	Interferences: OES libraries[†]	Calculated interferences[‡]
Ag	I R	328.068	3.7782	4.37	Ce Mn Rh	
Ag	I R	338.289	3.6640	2.23		
Al	II R	167.079	7.4208	4.64		
Al	II	172.127	11.8468	1.50		Sn Cr
Al	I r	237.839	5.2253	0.01	Hg Zr	V Fe Mn Zr
Al	I R	256.798	4.8267	0.17	Fe Mo Ru Zn	Cr
Al	I R	308.215	4.0215	0.82	OH	V Ta
Al	I r	309.271	4.0217	0.73	OH	V
Al	I r	309.284	4.0215	0.08	OH	V Ta
Al	I R	394.401	3.1427	0.67	Ce Re Ru	
Al	I r	396.152	3.1427	0.67	**Zr**	
Ar	II	137.005	25.8624	0.04	P	
Ar	II	157.500	21.3521	0.10	Fe Si	Si
Ar	II	264.960	21.4268	1.00		Mn Mo V W
Ar	I	345.497	15.2113	0.02		Nb V Ti
Ar	II	349.155	22.7732	3.51		Nb
Ar	II	355.949	23.1626	4.70	N_2	Co Cr Mo
Ar	II	357.665	23.0148	4.60	N_2	V W Zr
Ar	II	358.848	22.9489	4.67	N_2	Fe Mo Zr
Ar	I	404.441	14.6884	0.01	**Fe** Hf N_2 W Zr	**Fe** Mo Ti V Zr
Ar	II	410.391	22.5151	1.17		Cr Mo Nb Ta Ti
Ar	I	415.859	14.5290	0.03		Sc Zr
Ar	I	420.067	14.4992	0.03	Mo Ti	Fe Mo V W
Ar	II	487.986	19.6803	3.85		Cr Nb Ta

194

Element	State	Line (nm)	Upper energy (eV)	s_L ($\times 10^{-12}\ m^2$)	Interferences: OES libraries[†]	Calculated interferences[‡]
Ar	I	696.543	13.3280	0.35		V Zr
As	I R	189.043	6.5586	0.75	**Cr** Pd	**Cr** Fe **Mo**
As	I	200.334	7.5402	0.00		Co Mo Pt V Zr
As	I	234.984	6.5880	0.74	Mo Os	Mo Nb Ti W Zr
Au	I R	242.796	5.1050	3.42	Cl Pt Sr	Sr
Au	I R	267.594	4.6320	1.83	Co Fe Nb Rh Ta V W	Ni Ti W
B	I r	182.652	6.7900	0.27		Ir
B	I r	208.959	5.9336	0.06		**Cr Ni** Pt **W**
B	I r	249.677	4.9644	0.19	Re Sn Ta **W**	Co, Nb, Sc, **W**
Ba	II	230.425	5.9834	4.40	Co Ir Os	Co Mo Ru S Ti
Ba	II R	455.403	2.7218	11.32	Ce Cr	Fe Nb
Be	II R	313.041	3.9595	0.86	Ce N_2 Ta V W	Ta Ti V W
Be	I	332.108	6.4573	0.12	Be Ce Cr Pd Ru	Cr Mo Nb Ru Sc Ti V W Zr
Bi	I R	306.771	4.0405	1.87	Mo OH Sn	La Zr
Br	I R	148.861	8.3289	0.15	Cu Ru	
C	I r	156.140	7.9460	0.09	Tl	Cu Fe
C	I r	165.700	7.4879	0.17		Fe
C	I	193.093	7.6849	0.22		Cr Co Mo Sn Ta Ti
Ca	II R	393.366	3.1510	4.93	Ag Ce **Fe** Hf Ir	Ru Sc
Cd	I R	228.802	5.4172	8.60	As Ir Pt	**As** Fe Sc
Cd	I R	326.105	3.8009	0.02	Ce Ru V W	Ca Mo Pt Ru V W Zr
Cd	I	346.620	7.3769	3.76		Fe La Mo Nb Tc Ti V
Cd	I	361.051	7.3791	3.87	Fe Ni Re	Fe Mn Mo Ni V
Ce	II	413.765	3.5121	2.74	Os Re W	W
Cl	I R	118.884	10.4291	0.15		Ge Pt
Cl	I R	133.581	9.2817	0.05	**C**	**C**
Cl	I R	134.732	9.2024	0.26		Fe Se
Cl	I R	138.978	8.9212	0.34		Fe Pt Re
Cl	II	479.452	15.9606	3.91	La Mo Ru	Ca Fe La Mo Zr W
Co	II	228.615	5.8371	1.52	Ir	Nb Ni Y

Element	State	Line (nm)	Upper energy (eV)	s_L ($\times 10^{-12}$ m^2)	Interferences: OES libraries[†]	Calculated interferences[‡]
Co	I	340.511	4.0719	1.29	Bi Cr Ti V	Fe Mo Ti V W
Co	I	345.351	4.0209	1.78	**Cr** Re	**Cr** Mo Nb **Ti**
Co	I	351.835	6.9335	0.95		La Mo Nb V W
Co	I	387.311	6.8880	0.06	Ir Mn Ti	Mo Nb Ti
Cr	II	267.716	6.1792	1.59	Ce Mn P Pt Re Ru Te **W**	Pt V **W**
Cr	II	298.919	7.8856	1.81	Bi Os Ru Ta	Bi Ca La Hf Mo Ta V Zr
Cr	I R	425.433	2.9135	0.94	Bi Mo Nb	La Zr
Cr	I R	428.972	2.8895	0.55	Ti	La
Cs	I R	459.311	2.6986	0.04		Mo Sc Ta V
Cu	II	219.227	8.4865	1.40		Nb Sn V
Cu	I	219.959	7.0239	0.06		Cr Ga Mn Nb Ta Ti
Cu	II	224.700	8.2349	0.91		Pb Pt V W
Cu	I	296.116	5.5748	0.04	N_2	Mn Os Ru V W Zr
Cu	I R	324.754	3.8167	3.26	Ag Ce Fe Mn Mo Nb	Nb Zr
Cu	I R	327.395	3.7859	1.67	Ce **Co** Mo Nb	La **Ti** V
Cu	I	402.263	6.8673	0.84		Fe Cr Co Ge Mo Nb V
Cu	I	458.695	7.8047	0.79		Co Fe Mn Mo Sb Ta Ti V W Zr
Cu	I	510.554	3.8167	0.06		Cr Fe Mo Sc Ti Zr
Cu	I	515.323	6.1912	9.29	Ta	Al Fe Na Nb Ta
Cu	I	521.820	6.1921	8.69		Hf Nb V Zr
Eu	II R	372.493	3.3276	2.99	La Ru Ti	La Nb Rh Ru Ta
F	I	685.601	14.5048	3.91		Co Fe Mn Mo Pt Ti
F	I	690.246	14.5267	2.93	Ne	La Nb Ni Ti V Zr
Fe	I	208.972	7.5391	0.11		As B Co V
Fe	II R	238.204	5.2034	1.96		La Nb Pt
Fe	II r	239.563	5.2216	1.37	Cr Ni	Co Cr Mo V
Fe	II	249.326	7.6062	2.24		Mo Ti V W Zr

Element	State	Line (nm)	Upper energy (eV)	s_L ($\times 10^{-12}$ m^2)	Interferences: OES libraries[†]	Calculated interferences[‡]
Fe	I	249.400	5.9809	0.13	**Co** Ru Ti	Al **Co** Mn Nb Ta Os W
Fe	II R	259.940	4.7683	1.23	Ir Mo Ru Ta	Co Ta W
Fe	II	260.017	7.9197	0.05		Mn Nb Ta V W
Fe	II	271.441	5.5526	0.26		Nb Ti Zr
Fe	I	271.487	5.5237	0.23	Rh Ta V	Cr Ta Ti W
Fe	II	273.955	5.5108	1.24	Cr V	Co Cr Nb V W Zr
Fe	I R	302.064	4.1034	0.67		Cr Hf Mo Nb
Fe	I R	371.994	3.3320	0.32	Ce Os	Ba
Fe	I	373.486	4.1777	1.49		Mo Nb Ti
Fe	I R	385.991	3.2112	0.18	Ta W	Mo Sc Ta Zr
Ga	I r	294.364	4.3132	1.91		Nb V
Ga	I R	403.298	3.0734	1.14	Mn N_2 Ta Tb	Co Cr Mn Ta V Zr
Ga	I r	417.204	3.0734	1.19	Ce Fe Ti	
Gd	II r	376.840	3.3677	2.77	Cr Gd W	La Mo Ti W
Ge	I	303.907	4.9620	1.71	Ir	Cr Ta V Zr
H	I R	121.567	10.1989	0.06		
H	I	486.134	12.7486	0.07	Cr	Ca Cr Fe Hf Mn Mo Ru V W Zr
H	I	656.280	12.0876	0.75		Al Ba Ca Cr Mn Mo Ni Sc Ti W
Hf	I R	286.637	4.3242	2.06	V W	Mo W
Hg	I R	253.652	4.8865	0.24	Pt Rh	Co Cr Hf Ti
Hg	I	435.832	7.7305	2.80		Ca Fe Ir Mn Pb Re Sc Tc Zr
I	I R	145.798	8.5039	0.08	Cu	Co Cr Cu V
I	I R	183.038	6.7737	0.07		Mo
In	I	325.609	4.0810	3.07	Ce Fe Mn Mo	Pt W
In	I R	410.177	3.0219	1.76	Ce Ru	La Ru
In	I	451.131	3.0219	2.10	Ti Ru Ta	
Ir	I R	203.358	6.0949	0.34	Os P	Fe W
Ir	I	322.078	4.1999	0.38	Ce Hf Nb	La Mo Nb
Ir	I R	380.012	3.2617	1.65	Ce N_2 Ru Y	La Ta
K	I R	404.414	3.0649	0.04	N_2 W	Cr Mo Nb W
K	I R	404.721	3.0626	0.02	N_2	Cr Mo Nb W
K	I R	766.491	1.6171	9.36		

Element	State	Line (nm)	Upper energy (eV)	s_L ($\times 10^{-12}$ m^2)	Interferences: OES libraries[†]	Calculated interferences[‡]
La	II R	408.671	3.0330	1.92	Sc	Nb Sc Ti
Li	I R	323.266	3.8343	0.01	Os Ru Sb	Mo Nb W
Li	I	610.364	3.8786	2.72		
Li	I R	670.776	1.8479	2.44	N_2	
Mg	I	277.669	7.1755	1.20	W	Cr Ta Ti W
Mg	II	279.800	8.8637	3.31		Cr Hf Ir Mo Ni Re Ta W
Mg	II R	280.270	4.4225	1.20	Ce Co Cu Mn Ru Ti V	La Te Ti
Mg	I R	285.213	4.3458	7.16	Hf Ir Mo Zr	La
Mg	II	293.651	8.6548	0.61		Ca Co Fe Mo Nb Pt V Zn Zr W
Mg	I	383.230	5.9460	2.17		Cr Mo Nb Pd V W Y Zr
Mg	I	383.829	5.9460	2.49	**Mn** Ru Zr	Fe Mo V W Zr
Mn	II R	257.611	4.8115	2.36	Co Hg	
Mn	I	403.179	6.2093	0.18	Ce Fe La N_2 Ti V	Ca Co Fe La Mo Nb Rh Ta Ti V
Mn	I R	403.306	3.0734	0.35	Cr Ga N_2 Ta	Co Cr Ga Sc Sr Ta V Zr
Mn	I R	403.448	3.0723	0.22	Nb N_2	**Cr Fe**
Mo	I R	317.034	3.9097	1.86	Cl **Fe** Ta W	Nb
Mo	I R	379.825	3.2633	2.03	Nb N_2 Ru Ti	Nb
Mo	I R	386.410	3.2077	1.49	V	**Co Cr**
Mo	I R	390.295	3.1758	1.10	Cr Fe N_2	Ru
N	I	149.255	10.6904	0.11	Cu	Cr Fe Mn P V
N	I	174.267	10.6904	0.01	As	As Cr Ge Mn **Ni** Mo V
N	II	411.001	26.2125	0.77	Fe La V	Ca Cr Mn Mo Nb V Zr
Na	I R	330.237	3.7534	0.04	Bi **Cr** Re	Ag **Cr** Mo Ta W
Na	I R	588.995	2.1044	5.09	Cr Mo N_2	
Na	I R	589.592	2.1023	2.55	N_2	
Nb	II	313.078	4.3983	3.39	Ce N_2 Rh Ta Ti	Ta Ti V
Nb	II	316.340	4.2939	3.18	Ce Ta W	**Cr** Mo Ni Ta V **W**
Nb	I r	410.092	3.0711	2.87	Hf	Ta

Element	State	Line (nm)	Upper energy (eV)	s_L ($\times 10^{-12}$ m^2)	Interferences: OES libraries[†]	Calculated interferences[‡]
Nb	I r	416.466	3.0248	1.48	Pt	Ni Pt
Nd	II R	430.357	2.8802	2.40	Hf	W
Ni	II	225.385	6.8215	1.12	**Co** Hf Pb Re	**Co** Cr Fe Pb V W
Ni	I r	341.476	3.6552	0.92	K **Co** Ru **Zr**	Co Mo Nb Ta **Zr**
Ni	I r	349.296	3.6576	0.82	Ce **Mn**	**Cr Fe** Mo Nb V
Ni	I r	351.505	3.6353	0.83		Cr Mo Nb V
Ni	I r	352.454	3.5422	1.02	Mn V	Al Mo V W
Ni	I	361.939	3.8474	1.43		Mn Nb V Zr
Ni	I	460.036	6.2916	0.50		Co Cr Fe Mn Mo Nb Ta Ti W
O	I R	130.217	9.5214	0.08		S
O	I	777.196	10.7410	4.14		Ca Cr Co Cu La Mn Nb Fe V
Os	I R	330.157	3.7543	0.03	Re Ru Sr	Fe La Mo Sr
P	I R	177.495	6.9853	0.44	Au **Cu** Hf Pt	**Cu Mn Nb Ni W**
P	I R	178.283	6.9544	0.28	Na	I
P	I	185.890	8.0784	0.03		**Fe** Mo Nb Sb Se
P	I	253.561	7.2128	0.37	Fe Ta	Co Cr Fe Ru Ta Sr
Pb	II	220.356	7.3707	1.02	Nb Rh	**Co** Fe Nb W Zr
Pb	I	261.418	5.7109	3.47		Co Fe Mo Nb V
Pb	I	280.200	5.7441	3.02		Nb Ta
Pb	I R	283.305	4.3751	2.43	OH	Mo Ru
Pb	I	363.958	4.3751	1.01		Ba Co Nb Ta
Pb	I	368.348	4.3345	1.53		Ti V W Zr
Pb	I	405.783	4.3751	2.19	In Mg **Mn** N_2 Ti V Zn	Ba **Mn** Ti
Pd	I	340.458	4.4545	3.34	Ce Re	La Nb
Pd	I	360.955	4.3954	3.18	Ce Cr Ti	Mn Mo
Pd	I	363.469	4.2240	1.36	Fe	Nb
Pr	I R	433.392	2.8600	0.15	La N_2 V	Hf La Mo V
Pt	I R	265.944	4.6607	0.11	Pa Ru Ta	Nb P Ta
Pt	II	279.422	6.6798	0.06		Al Co La Mg Ti V W Zr
Pt	I R	306.471	4.0444	0.06	Hf Mo Ni OH Re Ru	Fe Mo Ni Tc Ti W Zr
Rb	I R	420.179	2.9499	0.11	Ar Fe Mn Ni	Fe Ta Ti V

Element	State	Line (nm)	Upper energy (eV)	s_L ($\times 10^{-12}$ m^2)	Interferences: OES libraries[†]	Calculated interferences[‡]
Re	I R	345.187	3.5908	0.53		Nb W
Rh	I R	343.489	3.6086	3.15		Sc
Rh	I	437.481	3.5389	0.72	Ca Co Mn Y	Mo Ta V Y Zr
Ru	I r	372.692	3.4734	1.82	Ce Fe Ir Re	Ta Ti
Ru	I R	372.803	3.3248	2.11	Ir	La W
S	I R	180.731	6.8602	0.32	N	Ni Mo **Sn**
S	I	189.408	9.2959	0.27	Fe	Fe Mo Sc
S	II	200.227	19.8505	0.73	Te	Co Cr Fe Mn Mo Nb Ni Si Ti
Sb	I R	206.834	5.9925	0.01	Ir Os	Co **Cr** Pt V W
Sb	I	252.852	6.1238	0.02	Mn Si V	Ba Co Cr Fe Mn Mo Si V W
Sb	I	259.804	5.8262	0.06	Mn Re	Cr Ni W
Sc	II	424.682	3.2337	2.84	Ce P Re Ru	C Mo
Se	I R	196.090	6.3229	0.01	Co Na Pd	Co Cr Cu Fe Pd
Si	I r	250.690	4.9538	0.28		Co Cr Mn Sc V
Si	I r	251.611	4.9538	0.43	Mo Re Ru V	Fe V
Si	I r	252.851	4.9297	0.17		Mo V W
Si	I	288.158	5.0824	0.62		Mn **Mo** Ta V **W Zr**
Si	I	390.552	5.0824	0.48	N$_2$	Ca Cr Fe Mo
Si	I	576.298	7.7700	0.10		Mn Ta V
Sm	II	360.428	3.9237	0.23	Fe Re Ti	Fe Mo Ti V Y Zr
Sm	II	411.855	3.6689	0.36	Fe Pt Ru	Fe Hf Pt Ru Sc Ta Ti V
Sm	II	443.432	3.1737	0.15		Hf Ta Ti Zr
Sn	II R	175.791	7.0530	0.38		P Sb Sc
Sn	II	189.990	7.0530	0.49		Sc Ti V
Sn	I	242.170	6.1861	2.31	Re	Fe Mn Re Ru V
Sn	I	283.999	4.7894	1.81	Cr Mn OH	Re W
Sn	I r	300.914	4.3288	0.48		Ca Mo Nb V W
Sn	I r	303.411	4.2949	0.86	**Cr** Ru W	**Cr** Mo Nb
Sn	I	317.505	4.3288	0.89	Ce **Co Fe** Te Ru	**Co** La V
Sn	I	326.233	4.8673	2.61	Fe Hf Pb Os	Mo V Zr
Sn	I	380.103	4.3288	0.43	N$_2$	Hf Mo Nb Pt Ti V
Sr	II R	407.771	3.0397	7.57	Cr Hg	
Sr	I R	460.733	2.6903	23.40		
Ta	II	239.993	5.5589	0.33	Hg Ru V	Cr Ru V
Ta	I R	301.254	4.1144	0.33		Nb Pt Zr

Element	State	Line (nm)	Upper energy (eV)	s_L ($\times 10^{-12}$ m^2)	Interferences: OES libraries[†]	Calculated interferences[‡]
Ta	I	301.637	5.5033	2.52	Fe Ir Mn Re W	Fe Mn Nb V W
Ta	II	302.017	5.9012	0.12		Ca Cr Fe Sc Si V W
Ta	I	362.661	3.9093	0.24	Rh	**Fe Mo** Rh Ru
Tb	II R	384.873	3.2205	0.19		Fe Ta
Tb	II R	387.417	3.1994	0.02	Co Ti V W	Cr Fe La Mo Pb Sc Ti V W Zr
Tb	II r	403.302	3.1994	0.22	Cr Ga Mn N_2 Ta	Co Fe Ga Mn Nb Sc Sr Ta V Zr
Tb	I R	432.643	2.8650	0.54	Nb N_2 Ti	Co Mo Nb Sr Ti
Te	I R	214.281	5.7843	0.02		Fe Mo Nb Re V W
Te	I R	225.903	5.4867	0.15	Ir Re	Cr Fe Ga Ir
Te	I	238.579	5.7843	0.04	Cr Ir	Cr Te
Ti	II	282.712	8.0709	0.83	OH	Al Ca Cr Fe Mo Nb Se Ta V W
Ti	II r	323.452	3.8809	1.70		Fe Ni
Ti	II r	334.941	3.7494	2.79	N_2	Cr Mo Nb W Zr
Ti	II r	337.280	3.6866	1.66	N_2 Pd Pt	Pd V **Zr**
Ti	I	360.105	5.7336	0.02		Ca Cr Fe La Mn Mo Sc Ti Zr
Ti	I r	365.350	3.4406	1.32	Ce Nb Re	**Ni**
Tl	I	351.923	4.4883	4.78		
Tl	I R	377.572	3.2828	2.07	Mo Ni N_2 V	
U	II r	385.957	3.2473	0.08		Cr Mo Mn Sc Ta Zr
V	II	309.311	4.3994	2.23	OH	Mg Ta W Zr
V	II	311.070	4.3329	2.00	Be Co Cr Fe Hf Mn Os Re Ru Zr	Co Mo Zr
V	I r	318.397	3.9330	3.07	Ti W	Mo Nb
V	I	411.177	3.3152	2.11	**Cr**	**Cr** Sc
V	I	437.923	3.1311	3.37	Bi Hf	Hf Mn Y
W	I	196.474	6.6764	0.12		Ca Ni Ta
W	II	200.810	6.7570	0.56		Co Sn
W	II	203.000	6.8678	0.48		Ca Cr Fe Mn Nb Re
W	I	400.875	3.4579	0.77	**Ti**	**Ti**
W	I	429.461	3.2521	0.39	Ce Cs Hf Ru Zr	Hf Ti Ru Zr

Element	State	Line (nm)	Upper energy (eV)	s_L ($\times 10^{-12}\ m^2$)	Interferences: OES libraries[†]	Calculated interferences[‡]
Y	II r	371.029	3.5205	3.96	Mo Nb	Mo Zr
Y	II r	377.433	3.4136	3.33	Os	La
Zn	II R	202.548	6.1193	1.89		Cr Fe
Zn	I R	213.857	5.7957	7.18	**Cu Ir Ni**	**Cu Ni** W
Zn	II	250.199	10.9649	0.65	Ru W	Fe Mo Ru Ta W
Zn	I	330.258	7.7828	2.48	Ta Zr	Ag **Ca** Cu **Nb** Ta Tc Ti
Zn	I	330.294	7.7824	0.83	Cr La Na Ta	Ca Cr Cu Ir La Na Nb Ta Ti V W
Zn	I	334.502	7.7834	3.07	Cr N_2	Ca Co Cr Fe Mo Nb Re Sc Ta
Zn	I	472.215	6.6546	0.83		As Co Cr Fe Mo Nb Sr Ta Ti V
Zn	I	481.053	6.6546	0.85	Cr Nb Rh	Cr Mo Nb Ti V W
Zr	II r	339.198	3.8182	4.00	Fe Mo Ru	Mo
Zr	I	350.592	4.1863	1.38		Mo Nb Sc W
Zr	I r	360.117	3.5958	0.39	Mn Ti	Ru

Note: When searching for lines likely to show non-linear regression due to self-absorption, these will normally be atomic resonance or near-resonance lines (*i.e.* I R or I r) with high s_L. A N_2 interference would normally indicate an air leak.

[†] Collected from: R.L. Kelly, A Table of Emission Lines in the Vacuum Ultraviolet for All Elements, CA, 1961; A.N. Zaïdel, *Tables of Spectral Lines*, Pergamon Press, New York, 1961; G.R. Harrison, *Wavelength Tables with Intensities in Arc, Spark, or Discharge Tube of more than 100,000 Spectrum Lines*, The M.I.T. Press, Cambridge, MA, 1969; A.R. Striganov and N.S. Sventitskii, Tables of Spectral Lines of Neutral and Ionized Atoms, IFI/Plenum, New York, 1968; and R.W.B. Pearse and A.G. Gaydon, *The Identification of Molecular Spectra*, 4th edn, Chapman and Hall, New York, 1976. Interferences in bold letters are reported in GDOES, mainly from Z. Weiss and K. Marshall, *Thin Solid Films* 1997, 308–9, 382 and an unpublished ATS Report, France, 1999.

[‡] Calculated using a 20 pm spectral window from: R. Payling and P.L. Larkins, *Optical Emission Lines of the Elements*, John Wiley & Sons, Chichester, 2000. Interferences in bold letters are reported in GDOES, from Z. Weiss and K. Marshall, *Thin Solid Films* 1997, 308–9, 382; and an unpublished ATS Report, France, 1999.

Appendix B. Additional Optical Emission Lines of Interest in GDOES[†]

Element	State	Line (nm)	Element	State	Line (nm)	Element	State	Line (nm)
Ag	II	224.642	Au	I	191.961	Cd	II R	214.439
Ag	II	241.318	Au	I	197.817	Cd	II R	226.502
Ag	II	243.779	Au	I	201.208	Ce	II r	393.108
Al	I R	214.556	Au	II	208.209	Ce	II	399.924
Al	I R	216.883	B	I R	136.246	Ce	II	413.380
Al	I r	220.462	B	I R	182.598	Ce	II	418.659
Al	I R	220.467	B	I R	208.889	Ce	II	428.993
Al	I r	257.509	B	I r	249.775	Ce	II	446.021
Al	I r	257.540	Ba	II	233.527	Cl	I r	136.353
Al	I R	265.248	Ba	II R	493.408	Cl	II	479.452
Al	I r	266.039	Ba	II	614.171	Cl	I	725.664
Al	II	281.619	Ba	II	649.690	Cl	I r	135.174
Al	I	305.007	Be	I R	234.861	Co	II	230.786
Al	I	305.714	Be	I	249.454	Co	II	231.160
Al	I	515.253	Be	I	249.458	Co	II	236.379
Ar	I	317.296	Be	I	249.473	Co	II	237.863
Ar	II	347.676	Be	I	265.045	Co	II	238.892
Ar	II	351.441	Be	II R	313.106	Co	I R	346.579
Ar	II	354.584	Bi	II	153.316	Co	I	352.981
Ar	II	356.433	Bi	II	190.230	Cr	II R	205.560
Ar	II	358.239	Bi	I R	195.470	Cr	II R	206.158
Ar	II	376.527	Bi	I R	206.163	Cr	II	283.563
Ar	II	378.086	Bi	I R	222.821	Cr	II	284.325
Ar	II	386.853	Bi	I R	223.061	Cr	I R	357.868
Ar	II	407.200	Bi	I R	145.004	Cr	I R	359.348
Ar	I	416.418	Br	I	153.190	Cr	I R	427.480
Ar	II	427.752	Br	I R	154.082	Cs	II	452.675
Ar	II	434.806	Br	I	163.357	Cs	I R	455.523
Ar	II	442.600	Br	II	470.486	Cs	I	672.326
Ar	II	460.955	Br	II	478.550	Cs	I	697.329
Ar	II	480.601	Br	II	481.671	Cs	I R	852.110
As	I R	159.360	C	I	199.365	Cu	II	213.598
As	I R	189.043	C	I	247.856	Cu	I	221.458
As	I R	193.760	C	II	336.175	Cu	I	221.566
As	I R	193.760	C	I	437.133	Cu	I R	222.571
As	I R	197.263	Ca	II	315.887	Cu	I	222.778
As	I R	197.263	Ca	II	317.933	Cu	I	223.009
As	I	228.811	Ca	II R	396.847	Cu	I	223.846
Au	II	174.047	Ca	I R	422.673	Cu	II	224.262

Element	State	Line (nm)	Element	State	Line (nm)	Element	State	Line (nm)
Cu	II	229.437	Fe	I	382.588	Ho	II r	341.644
Cu	I	261.837	Fe	I r	388.628	Ho	II R	345.602
Cu	I	276.637	Fe	I	404.581	Ho	II r	347.425
Cu	I	282.437	Fe	I	410.415	Ho	II R	381.074
Cu	I	674.142	Ga	II R	141.443	Ho	II r	389.094
Dy	II R	340.780	Ga	I r	250.019	Ho	II R	404.547
Dy	II R	353.171	Ga	I R	287.423	I	I R	142.549
Dy	II	353.602	Gd	II r	303.284	I	I R	161.761
Dy	II r	364.540	Gd	II r	310.050	I	I R	178.276
Dy	II R	394.468	Gd	II r	335.048	I	I	179.909
Dy	I	396.863	Gd	II r	335.863	I	I	184.445
Dy	II r	400.045	Gd	II r	336.224	I	I	206.163
Dy	I R	421.172	Gd	II r	342.246	I	II	516.122
Er	II r	323.058	Gd	II r	358.496	I	II	546.463
Er	II R	326.478	Gd	II r	364.620	In	II R	158.645
Er	II	333.386	Ge	II r	164.919	In	II R	230.615
Er	II R	337.275	Ge	I r	169.894	In	I R	303.936
Er	II r	349.910	Ge	I r	206.866	In	I	325.856
Er	II r	369.265	Ge	I r	209.426	Ir	I	204.323
Er	II R	390.631	Ge	I	219.871	Ir	I	205.222
Eu	II R	381.967	Ge	I r	259.253	Ir	I R	208.883
Eu	II r	390.711	Ge	I r	265.117	Ir	I	212.642
Eu	II r	393.050	Ge	I R	265.157	Ir	I	215.265
Eu	II R	412.973	H	I	410.176	Ir	I	224.276
Eu	II R	420.504	H	I	434.049	Ir	I	236.805
Eu	I R	459.404	H	I	656.280	Ir	I	254.397
F	II	156.478	He	II	468.570	Ir	I R	284.978
Fe	II R	234.350	Hf	II R	232.248	Ir	I R	292.479
Fe	II r	240.489	Hf	II	251.689	Ir	I	343.701
Fe	II r	259.837	Hf	II R	263.872	Ir	I R	351.365
Fe	II	259.879	Hf	II	264.141	K	I R	769.897
Fe	II R	259.940	Hf	II	273.877	Kr	I	557.029
Fe	I r	302.049	Hf	II	277.336	Kr	I	587.091
Fe	I R	344.061	Hf	II	282.023	La	II	333.749
Fe	I	355.951	Hf	I R	286.637	La	II r	379.477
Fe	I	357.024	Hf	II R	339.979	La	II	394.910
Fe	I	357.676	Hg	II R	164.995	La	II	398.852
Fe	I	358.119	Hg	I R	184.950	La	II r	407.734
Fe	I	358.861	Hg	II R	194.231	La	II	412.322
Fe	I r	374.826	Hg	I	296.728	La	II r	433.375
Fe	I	374.949	Hg	I	365.015	La	I	624.991
Fe	I	375.823	Hg	I	404.656	Li	I R	274.120
Fe	I	382.043	Ho	II R	339.895	Li	I	460.289

Element	State	Line (nm)	Element	State	Line (nm)	Element	State	Line (nm)
Li	I R	670.776	N	I	409.880	Ni	II	225.783
Lu	II R	219.556	N	II	410.426	Ni	II	226.446
Lu	II R	261.541	N	II	566.663	Ni	II	227.021
Lu	II	289.484	N	II	567.601	Ni	II	227.473
Lu	II	291.139	N	II	567.956	Ni	II	227.568
Lu	II	307.761	N	I	746.860	Ni	II	227.667
Lu	II	339.707	Na	I R	285.301	Ni	II	227.832
Lu	II	347.248	Na	I R	330.298	Ni	II	228.708
Lu	II	355.442	Na	I	568.820	Ni	II	230.299
Lu	I R	451.857	Na	I R	588.995	Ni	II	231.604
Mg	I R	202.582	Nb	II r	269.706	Ni	I R	232.003
Mg	I	267.246	Nb	II	309.417	Ni	II	256.637
Mg	I	278.142	Nb	II	319.497	Ni	I R	336.956
Mg	II	279.078	Nb	II	322.548	Ni	I	338.057
Mg	II R	279.553	Nb	I r	405.893	Ni	I r	339.299
Mg	II	292.863	Nb	I r	407.972	Ni	I r	342.371
Mg	I	294.199	Nb	I r	412.381	Ni	I r	343.356
Mg	I	332.992	Nb	I R	413.709	Ni	I r	344.626
Mg	I	333.215	Nd	II	378.425	Ni	I r	345.289
Mg	I	333.667	Nd	II	401.224	Ni	I r	345.846
Mg	I	382.936	Nd	II	406.108	Ni	I r	346.165
Mg	I	383.230	Nd	II	410.945	Ni	I r	347.255
Mg	I	516.732	Nd	II r	415.607	Ni	I r	359.770
Mg	I	517.268	Nd	II r	417.732	Ni	I r	361.046
Mg	I	518.361	Ne	I	540.056	Ni	I	460.622
Mg	I	552.841	Ne	I	585.248	O	II	486.097
Mn	II R	259.372	Ne	I	640.224	O	I	615.819
Mn	II R	260.568	Ni	I r	215.831	O	I	777.418
Mn	I R	279.482	Ni	II	216.555	O	I	777.540
Mn	I R	279.827	Ni	II	216.909	Os	I	189.834
Mn	II	293.306	Ni	II	217.467	Os	I R	190.099
Mn	II	293.931	Ni	II	217.514	Os	II	206.723
Mn	II	294.921	Ni	II	218.047	Os	II R	225.585
Mn	I R	403.075	Ni	II	218.460	Os	II R	228.228
Mo	II R	202.032	Ni	II	218.550	Os	II	233.681
Mo	II R	203.846	Ni	I r	219.735	Os	I R	290.906
Mo	II R	204.598	Ni	II	220.141	Os	I R	305.865
Mo	II	281.615	Ni	II	220.671	Os	I	326.229
Mo	II	284.824	Ni	II	221.038	Os	I R	326.795
Mo	II	287.151	Ni	II	221.648	Os	I R	442.047
Mo	I R	313.260	Ni	II	222.040	P	I R	138.148
N	I	149.481	Ni	II	222.295	P	I	168.598
N	I	174.546	Ni	II	222.486	P	I	169.403

Element	State	Line (nm)	Element	State	Line (nm)	Element	State	Line (nm)
P	I R	178.765	Pt	II	203.647	S	I r	182.625
P	I	185.789	Pt	II	204.917	S	I	469.411
P	I	185.887	Pt	II	214.425	S	I	469.544
P	I	185.941	Pt	I	217.647	S	I	469.625
P	I	213.618	Pt	I R	283.029	Sb	I R	206.834
P	I	214.914	Pt	I R	292.980	Sb	I R	217.582
Pb	II R	168.214	Pt	I r	299.797	Sb	I R	231.146
Pb	I R	168.791	Pu	II	363.224	Sb	I	287.791
Pb	II	182.206	Pu	II	402.154	Sb	I	323.253
Pb	I R	217.000	Pu	II	453.615	Sb	I	326.751
Pb	I	223.743	Ra	II R	381.442	Sc	II	335.372
Pb	I	239.380	Ra	II R	468.227	Sc	II r	357.253
Pb	I	240.195	Ra	I R	482.591	Sc	II r	361.383
Pb	I	241.174	Rb	I R	421.552	Sc	II r	363.074
Pb	I	244.384	Rb	I R	780.026	Sc	II R	364.278
Pb	I	244.619	Rb	I R	794.760	Sc	I R	390.749
Pb	I	247.638	Re	II	189.774	Sc	I r	391.182
Pb	I	261.418	Re	II R	189.836	Sc	I R	402.039
Pb	I	261.365	Re	II	197.279	Sc	I r	402.368
Pb	I	262.828	Re	II R	197.313	Sc	I	567.182
Pb	I	266.316	Re	I R	204.910	Se	I	185.518
Pb	I	282.320	Re	II R	221.427	Se	I	199.511
Pb	I	357.275	Re	II R	227.525	Se	I r	203.985
Pb	I	367.151	Re	I R	229.448	Se	I	206.279
Pb	I	373.994	Re	I R	346.045	Se	I R	207.479
Pb	II	560.886	Re	I R	488.914	Se	I r	216.417
Pd	II	229.653	Rh	II	233.477	Se	I	241.351
Pd	II	248.892	Rh	II	249.078	Se	I	254.796
Pd	II	249.880	Rh	I r	250.465	Se	I	473.075
Pd	II	265.875	Rh	II	252.052	Se	I	473.900
Pd	II	285.458	Rh	I r	332.309	Se	I	474.222
Pd	I	324.270	Rh	I R	339.682	Si	I r	198.899
Pd	I	342.122	Rh	I r	365.799	Si	I	212.412
Pd	I	351.694	Rh	I R	369.236	Si	I R	220.798
Pr	II R	390.843	Ru	II	240.273	Si	I r	221.089
Pr	II	406.281	Ru	II	245.644	Si	I r	221.175
Pr	II	410.072	Ru	II	245.658	Si	I r	221.667
Pr	II	414.311	Ru	II	266.162	Si	I	243.515
Pr	II r	417.939	Ru	II	267.878	Si	I R	251.432
Pr	II	418.948	Ru	I	269.214	Si	I r	251.920
Pr	II r	422.293	Ru	I R	349.894	Si	I r	252.411
Pr	II R	422.532	S	I	166.669	Si	I	253.238
Pt	II	177.708	S	I r	182.034	Si	I	263.128

Element	State	Line (nm)	Element	State	Line (nm)	Element	State	Line (nm)
Si	I	298.764	Ta	II	233.218	Ti	I r	392.453
Si	II R	152.671	Ta	I R	238.709	Ti	I R	394.867
Sm	II r	330.632	Ta	II	240.086	Ti	I r	395.634
Sm	II	356.827	Ta	II r	263.558	Ti	I r	395.821
Sm	II	359.260	Ta	II R	268.515	Ti	I R	396.285
Sm	II	360.949	Ta	II	269.451	Ti	I r	396.427
Sm	II r	363.427	Ta	I	331.114	Ti	I R	398.176
Sm	II	422.536	Ta	I	331.885	Ti	I r	399.864
Sm	II	428.079	Ta	II	340.667	Ti	I	498.173
Sm	I	429.674	Tb	II	321.998	Ti	I	499.107
Sm	II	442.434	Tb	I	332.443	Ti	I	499.950
Sn	II	147.501	Tb	II R	350.914	Ti	I	500.721
Sn	II	181.120	Tb	II	356.170	Ti	I	501.428
Sn	II	189.990	Tb	II r	367.636	Tl	II R	190.865
Sn	II R	215.152	Tb	II r	370.285	Tl	I R	237.958
Sn	I	226.892	Tc	I	371.886	Tl	I R	276.789
Sn	I r	235.484	Te	I R	214.281	Tl	I	291.832
Sn	I	242.170	Te	I	238.329	Tl	I	322.975
Sn	I	242.949	Te	I	253.076	Tl	I	535.046
Sn	I R	254.655	Te	I	276.973	Tm	II R	313.126
Sn	I	257.158	Th	II	274.716	Tm	II r	317.283
Sn	I r	266.124	Th		283.231	Tm	II r	336.262
Sn	I	277.981	Th	II	283.730	Tm	II r	342.508
Sn	I	278.502	Th	II	294.286	Tm	II R	346.220
Sn	I	281.257	Th		318.020	Tm	II R	376.133
Sn	I	285.060	Th		329.059	Tm	II R	376.191
Sn	I R	286.332	Th		332.513	Tm	II	384.806
Sn	I	333.062	Th		353.875	U	II R	263.553
Sn	I	452.475	Th		360.104	U	II r	279.393
Sn	II	645.341	Th		401.914	U	II	355.217
Sn	II	684.349	Th	II	438.186	U	II r	367.007
Sr	II	215.283	Ti	II r	190.821	U	II	367.258
Sr	II	216.591	Ti	II	275.161	U	II r	393.202
Sr	II	338.071	Ti	II	281.782	U	I	409.013
Sr	II	346.445	Ti	II r	323.658	U	I	424.164
Sr	II	347.489	Ti	II r	323.904	V	II r	268.796
Sr	II R	421.552	Ti	I R	334.188	V	II	290.881
Sr	II	430.544	Ti	II	334.904	V	II	292.401
Sr	I	483.204	Ti	II r	336.122	V	II	292.464
Sr	I	487.249	Ti	I	337.786	V	II	310.229
Sr	I	496.226	Ti	II R	338.377	V	II	311.837
Ta	II	226.230	Ti	I R	363.546	V	II	312.528
Ta	II	228.916	Ti	I r	364.268	V	I r	318.341

Element	State	Line (nm)	Element	State	Line (nm)	Element	State	Line (nm)
V	I r	318.538	Yb	II R	222.447	Zn	I	303.578
V	I	438.471	Yb	II R	289.139	Zn	I	307.206
V	I	438.998	Yb	II R	297.056	Zn	I R	307.590
W	II	207.912	Yb	II R	328.937	Zn	I	328.233
W	II r	209.475	Yb	II R	369.419	Zn	I	334.557
W	II r	218.936	Yb	I R	398.799	Zn	I	334.594
W	II	220.449	Zn	II R	206.200	Zn	I	377.438
W	II R	224.876	Zn	II	206.423	Zn	I	468.013
W	II	239.708	Zn	II	209.994	Zn	II	492.401
W	II	248.924	Zn	II	210.218	Zn	I	636.234
W	II	258.917	Zn	I	247.942	Zr	II r	257.147
W	I	321.556	Zn	I	249.140	Zr	II r	327.307
W	II	361.380	Zn	II	255.795	Zr	II r	327.927
W	I	430.210	Zn	II	256.780	Zr	II	343.714
Xe	I	450.097	Zn	I	256.987	Zr	II r	349.619
Xe	I	462.427	Zn	I	258.243	Zr	I R	351.959
Xe	I	467.122	Zn	I	260.856	Zr	I r	354.768
Y	II r	321.668	Zn	I	267.053	Zr	II R	357.247
Y	II r	324.227	Zn	I	268.416	Zr	I	468.781
Y	II r	360.073	Zn	I	275.645	Zr	I	471.008
Y	II R	363.312	Zn	I	277.085	Zr	I	473.948
Y	II r	371.029	Zn	I	277.097	Zr	I	477.232
Y	II r	378.870	Zn	I	280.086	Zr	I	481.562
Y	II	437.493	Zn	II	280.194			
Yb	II R	211.668	Zn	I	301.836			

† Collected from various GDOES and ICP spectra, and then compared with the calculated spectra from R. Payling and P. L. Larkins, *Optical Emission Lines of the Elements*, John Wiley & Sons, Chischester, 2000.
Note: Wavelengths shown are for vacuum below 200 nm and standard air above 200 nm.

Subject Index

www.ingramcontent.com/pod-product-compliance
Lightning Source LLC
Chambersburg PA
CBHW031950180326
41458CB00006B/1682